Ship of Dreams

Charles T. Whipple

SHIP OF DREAMS
Copyright © 2018 Charles T. Whipple
ISBN-13: 978-1729612170
ISBN-10: 1729612172

Cover photo provided by Charles T. Whipple
Cover Design & Interior Layout by Laura Shinn Designs
http://laurashinn.yolasite.com
Anchor image Designed & Created by:
    Macrovector / Freepik com

This book is a work of non-fiction.
This work is protected by both United States and International copyright laws and is provided solely for entertainment purposes. All rights reserved. No part of this book may be reproduced or transmitted in any form or by any means, electronic or mechanical, including photocopying, recording, or by any information storage and retrieval system, without permission in writing from the publisher. The scanning, uploading, and distribution of this book or any part thereof via the Internet or World Wide Web or by any other means without the expressed permission of the publisher is illegal and punishable by law.

Table of Contents

Acknowledgements	1
Forward	3
Chapter One	8
Chapter Two	19
Chapter Three	29
Chapter Four	42
Chapter Five	51
Chapter Six	61
Chapter Seven	70
Chapter Eight	82
Chapter Nine	93
Chapter Ten	104
Chapter Eleven	115
Chapter Twelve	127
Chapter Thirteen	141
Chapter Fourteen	150
Chapter Fifteen	159
Chapter Sixteen	170
Chapter Seventeen	182
Chapter Eighteen	194
Epilogue	205
About the Author	214

Acknowledgements

Certainly I will miss someone whom I should acknowledge, but here goes. John Welsford is at the top of the list, of course, as he drew all the lines and wrote all the words on vellum that enabled me to build my ship of dreams. His partner Denny at the top of the list, too, for unfailing hospitality, then and now. The members of WOS, who welcomed a crazy Yank into their midst, even though he was not a fellow naturalist, and who came in force to see me off when I left Tauranga Bridge Marina, bound for Hawaii.

Friends in New Zealand—Fran and Mik Borsos and the crew at Fran's Café for always having a slice of carrot cake in the cooler case just for me. Kiwi friends are too many to mention one by one, but too precious not to. Janice, who christened my ship of dreams; Jenny, who chairs Tauranga Writers, Inc.;

Kirsten, Paul, PJ, Bruce, Blair, and so many more who helped me make New Zealand my second home. Would be that I were a Kiwi, too.

Fred Jeanes and the crew at Tauranga Bridge Marina who welcomed me as a liveaboard and sounded the horn when I motored out of the marina toward the mouth of Tauranga harbor.

I must mention the helicopter crew that snatched me from the top of a fang of volcanic rock in the early morning hours. Lance Donnelly piloted the rescue chopper out of Mechanics Bay in Auckland with crewmen Leon Ford and paramedic Bruce Kerr. They did their jobs faultlessly and I am here writing these words, partly because of them.

George "Tab" Melton and Joshua Shinn went through the final manuscript to find all the mistakes and logical inconsistencies. Thanks, guys.

Finally there's the girl who stayed in Japan but let me pursue the dream I'd put off for so long (even though it changed over the years).

My thanks to all,
Charles T. Whipple

Foreword

    I get a lot of emails and letters from people with dreams of building their own little ship and setting off on a voyage to paradise. Very, very few of those actually progress any further than just the dream. Those fantasies, though, may help the dreamer cope with the extraordinary stresses of our ordinary day-to-day struggles with jobs, mortgages, and keeping up with others' expectations.

But the book you're holding (or scrolling through, as the case may be) tells the tale of a man who achieved and lived more of the dream than anyone I've ever encountered.

I had the dreamers in mind when I designed Swaggie, about the smallest, most achievable blue-water voyager that I considered to be practical: Practical in that a determined novice could build her, practical in that she'd have room for two very close friends for a month or so, practical in that she could carry enough stores for that time, and practical in that she'd survive the sort of weather one might encounter on that month at sea.

But back to the dreamer in question. An email showed up in my inbox one day, asking if Swaggie could be modified a bit here and there, changes to the interior, the rig, a "proper cockpit" (not needed in the original as being junk rigged, Swaggie was to be sailed from within the shelter of her small raised cabin).

I asked, "Where do you plan to voyage?"

"I want to come to New Zealand and build a boat in which I will circumnavigate."

Wow!

A couple more exchanges ensued, clarifying things, me persisting in spite of thinking that this was the absolute classic armchair sailor's fantasy, and that I was dealing with a "dreamer" here.

The plan was to sail eastward along the top edges of the "roaring 40s." Using the southern oceans' prevailing westerly winds to take the little ship across the widest stretch of ocean on the planet, from New Zealand to Cape Horn, thence to the Falkland Islands.

As the longest leg of the planned voyage, that set the criteria for stores. Food, water, fuel for cooking and engine, spares. Everything from toothpaste and soap to food and water. Even at the most careful use, and, with a reasonable reserve, that would weigh about a ton.

Now, a conventional sailing boat can carry about a quarter of its designed displacement in "variable load," that being the difference between full stores and water and empty, so we were looking at about a 3-ton boat plus consumables.

I drew a sketch of a little cutter that would carry that weight and not be unstable when empty, and which fit all the other items on the wish list, sent it off, and within a very short time had an email that told me when he would arrive!

Happens that would be about a week before the family and I were due to fly out for a month-long trip to the USA, for the most part out of communication!

I met Charles Whipple sitting on the seats in the arrival area of Auckland International Airport. I'd managed to screw up the arrival date, and did not have a cell phone, so I'd headed home, where I knew he'd be able to get a message to me, and gone back to pick him up. I think he's forgiven me for that, but we're careful now with flight details and timing.

At that time I was busy building a house for Denny, the family, and myself, and it was taking too long, so my time at the drawing board was limited. That plus we were about to head on out, so once we got Charlie settled in his little caravan at the Waikato Outdoor Society, where we were living while we waited for our house to become habitable (Charlie will have lots more to tell you about WOS). I drew several sheets of plans for things like hatches, the building platform, cambered laminated beams, and so on, enough, I thought, to keep him busy until we got back and I could get the drawings a bit ahead of him again.

My workshop has a good selection of machinery and tools, a bench saw, planer, jointer, drill press, woodturning lathe, and so on, a big collection of power and hand tools, some very specialized. I'm loath to allow anyone to work in "my space" and very possessive—those tools are like extensions to my hands—and it was with much trepidation that I left for our holiday with our guest working in there.

But it went OK, and by the time we got back he'd prefabricated the quite complex double coaming hatches, done a lot of laminating, figured out how to use Trade Me, New Zealand's equivalent to eBay, to find second-hand bargains for his project, and had made some friends among the members of WOS.

I'd based my time estimate on how long it would take a boatbuilder, experienced and skilled, to build the boat if working full time without any distractions. Perhaps I didn't explain that as clearly as I might have, or perhaps he only heard "six months" part, but it was clear early on that our guest was working at his writing job, learning many skills as he went, and that sometimes the "thinking chair" (a sawhorse if I remember rightly) were all soaking up some hours.

So progress was a lot slower than the hypothetical boatbuilder might have achieved. But that said, Charlie soldiered on, and there was something to see at the end of every day, well, almost anyway, there were a couple of trips home, which slowed him somewhat. As did the bench saw! We always hoped that our friend would return. He always did, got back to work, and progress continued.

It was wonderful to be a part of the build. I was careful not to intrude on what was Charlie's build, but did pitch in to do some of the more technically demanding things, one being fairing up stringers in the bow area so the planking would lie fair. Another was to shape that huge chunk of laminated kwila hardwood around the propeller aperture with a large angle grinder sporting a seriously dangerous woodcarving disc. I deemed it too dangerous for a novice (those things bite) so that was my task.

It was great too, to host the many visitors who came by to see the boat, to watch progress, and sometimes, like retired boatbuilder Hugh Green, who arrived just as we were about to epoxy the Kevlar crash-resistant covering onto the forward underwater sections, and then fiberglass over that and the rest of the hull. Andrew, a WOS member, who came and helped us roll the hull over, and who shared the special moment when she was upright for the first time, and watched as Charlie stood inside the interior of his little ship for the first time and marveled at how much space there was in her 21½ feet.

There were others—help to roll her over, help to fit the transom, help to drill and face bronze castings—so many people helped, mostly in small ways, but there were a few who really made a difference. All of them shared a little of Charles Whipple's dream, and became part of it.

I myself shared that dream. It lived within me, became part of me. I watched, helped out some where needed, sharpened tools, lifted things, went and drew more plans, and shared my precious workshop with my friend.

Sailing the little ship, so recently only a bunch of neurons firing in my fevered imagination, then drawn out on Mylar for Charlie to translate into 3-D reality, was beyond special. Launching day is one designers both love and dread, but she sat right on her marks, and once rigged and the sails bent on, she confirmed my theories about shape and proportions. Sailing on her has given me some very treasured memories.

A small community of friends came together around that project—my partner Denny and I, our son, our friends, Charlie's new friends at WOS, and others. Fran at Frans Café being a special one, and we did go and visit her when he came back to joggle his memory before writing this book. She even gave him the recipe for her famous carrot cake! A secret rarely shared and a real privilege!

The building of that little voyager was a special thing, so much more than just a man working in a shed. The project, the boat, the voyage became the focus around which so many relationships were built, and the boat herself gradually developed a personality, that of a short, somewhat rounded, but very sassy woman, the sort who'd take no nonsense from anyone, but who had a cheeky grin when she wasn't cussing you out, would do anything she could for you, and as we later found out, could dance the feet off anyone.

Her loss broke a lot of hearts, that of her builder more than anyone. But unlike so many of the dreamers I've dealt with over so many years, Charles Whipple did much more than dream. Much more.

ValeResolution, in our hearts you'll live forever.

John Welsford
June 2018

Chapter One

Recrudescence

More than 10 hours in the air from Narita, Japan's biggest door to the world, put me in the skies over *Aotearoa*. I'd forgotten how green this country is. My heartbeat quickened. New Zealand is my country of dreams, and I'd returned for some refresher courses in dreaming.

The Air New Zealand 787 I was on executed a near-perfect landing, then taxied flawlessly to its arrival gate.

New Zealand. *Aotearoa*. They call Australia the "down under" country, but New Zealand is farther south than Ozzieville. No motionless North Star in New Zealand. Only the warm, open arms of the Southern Cross. And beneath the Southern Cross lies the battered and broken remains of my ship of dreams. Her name was *Resolution*, taken from the ship James Cook used for his second voyage of discovery, which began in 1772, and for his third.

Resolution is renowned for her voyages to Hawaii and points north and east, but she also spent time in New Zealand, where I decided to build my own ship of dreams.

The 787 swooped down over green fields and the myriad islands that dot the Hauraki Gulf. Nine o'clock in the morning, and craft were already on the move, little pointy blocks with fan-shaped wakes spreading out behind them. Every sheltered cove is dotted with pleasure boats, power and sail, giving Auckland its motto—the City of Sails. For the city has more pleasure boats per capita than any other city on Earth. No wonder the country sets the pace when it comes to technical advances in sailing – as proved by its record in America's Cup competition.

Going through New Zealand immigration and baggage

inspection took only as long as it took me to walk the distance. I marked outdoor clothing on my customs form. Asked what, I lifted my foot and pointed at my walking shoes.

"Been hiking with those?" the officer said. She had a stern but friendly face, and I could see that she'd brook no nonsense.

"No. Just the walking path near my home."

The officer waved me on.

John Welsford waited in the arrival lobby. Minutes later, I got my introduction to Indy, the truck watchdog. The truck is a Mitsubishi Triton, a four-door with a cover over the pickup bed (no extra space there, as John keeps a full array of tools and whatnot in the back). I put my suitcase on the back seat along with my super-heavy shoulder bag.

I opened the front door, passenger side. Indy sat up.

Being a dog lover, I held my hand out so Indy could have a sniff and categorize me as harmless.

His ears went back, his teeth bared, and he snarled at me. I quickly jerked my hand away.

"Indy's the truck guard dog," John said. "Here, let me introduce you to him."

After I was formally introduced to the guardian of the truck, Indy and I shared the passenger side, him and my feet in the footwell and me on the seat.

For the first time in 10 years, I was back in New Zealand. *Aotearoa*. The Land of the Long White Cloud. A land of islands surrounded by oceans . . . clean, clear oceans. Great waters that resurrect voyaging dreams in the hearts of those who have come to think their dreams are long dead.

In the Beginning

Frustration makes a man start dreaming, imagining situations far removed from day-to-day concerns. In my case, the reasons for frustration were many and varied, but the dreams were consistent: Owning an ocean-going sailboat and voyaging freely from island to island, keeping a log and using it as the basis for article after article, book after book, on voyaging from hither to yon, and on traipsing across many an unfamiliar countryside and through many a quaint and undiscovered village filled with friendly, helpful people.

Well, that was how it went in the dream, anyway.

My home state is Arizona, a place with hardly enough water to float a toothpick, much less an ocean-going yacht. Still, my boyish daydreams often turned to clipper ships and windjammers. You see, my educator parents bought me a membership in the Landmark Book Club, and I received a new biography about a famous American every single month. The club opened my eyes and imagination to James Butler Hickok, better known as Wild Bill, Andy "Old Hickory" Jackson, Davy Crockett, Sam Houston, and perhaps most important to my dream, Angus McKay. Although I read the biographies of the men who pioneered the westward expansion of the United States, what really inspired me were the stories of the swift windjammers designed by Angus McKay and the new records they set, voyage after voyage. I soon learned to draw those windjammers, with sails billowing and a bone in their teeth, striving to voyage across trackless oceans in ever-shorter times.

Ship of Dreams

When I was in the third grade, I remember using white chalk on a blackboard to depict a sailing ship under full load of canvas while my classmates were outside for recess. I don't remember anyone's reaction, but I'll never forget that ship. In fact, when I was a senior in high school, our prom theme was "Red Sails in the Sunset," and I drew a windjammer big enough to cover the entire back wall of the dance hall.

Dreams

In 1974, my dreams took me to Hawaii with the girl who'd said she'd like to sail along with me, voyaging among the islands. I quit a good job in advertising and left a wife and two boys to follow that dream . . . .

On the plane to Hawaii, I said to the girl, "I don't know what kind of job I can find in Hawaii. I might be digging ditches. But I'll keep looking for the right boat, and when I find it, we can start our cruising life."

She nodded.

In Hawaii, I found a minimum-wage job at a very small ad agency, and another at the same rate as the 8-to-midnight disc jockey at Radio KOHO in Honolulu. The dream would not die, and every weekend I prowled the docks at Ala Wai, looking for the right boat.

I decided advertising account service was not my game. I signed up for a correspondence course from Westlawn School of Yacht Design (now Westlawn Institute of Marine Technology). At the same time, I took courses in writing from the Institute of Children's Literature, from which I eventually received a master's certificate.

I learned a lot about ocean-going yachts from Westlawn, but writing went better and quicker than yacht design for me. In 1975, I sold my first feature article to a magazine called *Dog Fancy*. That started me on the writing trail, but did not knock me off the search for the boat of my dreams. That same year, my first daughter was born at Kapiolani Hospital in Honolulu. If anything, her birth added impetus to my search for the perfect boat.

About that same time, I applied for a job at Hawaii's largest ad agency, and my boss found out. He fired me, which turned out to be a good thing because I ended up at the *Waikiki*

Beach Press, a Scripps League newspaper. My most significant writing assignment at the *Beach Press* was covering the return of the *Hokule'a*, a Herb Kane-designed traditional Hawaiian voyaging canoe. The *Hokule'a* had sailed to Tahiti and was then arriving back to Honolulu. My story ran a whole page in the newspaper. Since then, the *Hokule'a* has sailed around the world, and is still sailing.

By getting fired, I achieved two things. I gained a writing position and I met the president of Obun Printing, a Tokyo company that subcontracted with Dentsu, which was quickly growing into the world's largest advertising agency. The writing position affirmed me as a bona fide writer and Obun Printing later became my employer in Japan. But neither of those things pushed me further toward becoming an ocean voyager.

In 1976, an ad I wrote for the *Waikiki Beach Press's* Japanese edition won the top award in *Editor and Publisher* magazine's annual contest, and my KOHO radio program, "Midnight Express," was the most listened to nighttime program in Honolulu. But I was still looking for just the right boat.

Then I found her. She was cold-molded with a three-quarter length keel, and the owner wanted a mere $10,000. I took the girl who'd said she wanted to sail away with me, along with our daughter, to see the boat. A nice boat she was, too, all 32 feet of her.

After the agent showed us the boat, we sat in the main cabin.

"What do you think of the boat?" I said.

"It's nice," she said. Baby daughter slept in her arms.

"We can't afford to live in an apartment and pay for the boat at the same time," I said. "We'll have to live aboard. Can you do that?"

A look of incredulity swept across her face. "You're not really serious about sailing away, are you?"

The dream crashed.

Oh, I owned boats over the years and even did some long-distance offshore sailing. But as of that moment, voyaging as a way of life could never be a reality.

The boat I'd named *Resolution* in my mind went to another buyer, and I turned to being a father and a writer.

In January 1977, we returned to Japan and made our home there. Over three decades, I owned several boats: *Miss America*, a 24-foot sharpie schooner I designed and built myself; *Charlie's Angel*, a 16-foot unsinkable cutter in which I sailed the islands south of Tokyo; *Umisaurus*, a 34-foot center cockpit sloop; *Millennium Rhyme*, a vintage Bluewater 21; and *DoriKam*, a Westsail 32 that took me from Olympia, Washington, to La Paz in the Mexican state of Baja California Sur.

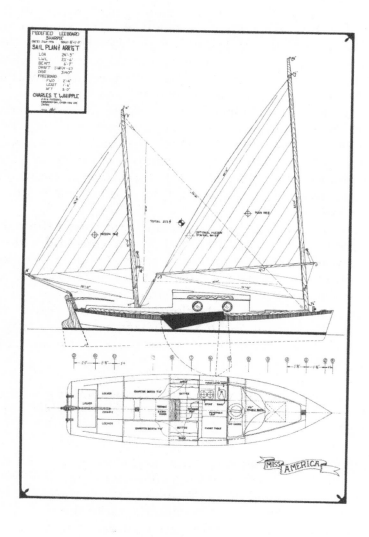

My friend Minoru Saito finished his seventh circumnavigation on June 6th, 2005, the oldest single-hander ever at age 71. That gave me confidence. I turned 64 that November and I planned to sail alone around the world.

Of course, other seniors have sailed around the world. Sir Francis Chichester comes to mind, along with David Clark. But these men sailed big boats. Saito's *Shuten-dohji II* is 50 feet long. Chichester's *Gypsy Moth* measured just over 53 feet, and Clark's boats were 44 and 34 feet long.

Still, some single-handed circumnavigators sailed small boats. John Guzzwell's *Trekka* measured 20'6". Hiroshi Aoki's *Ahodori* was 20'9" LOA, and Robin Lee Graham's *Dove*, 26 feet. It could be done. I planned to circumnavigate in a 21-foot blue-water voyager designed by John Welsford of Hamilton, New Zealand. Her name was *Resolution*.

John started this all for me with a small blue-water cruiser he called Swaggie. Eighteen feet long and very robust, she looked like the kind of boat in which to cross oceans. As a writer, I suggested to John that perhaps I could go to New Zealand, build a *Swaggie*, chronicle the process, and produce a book on how to build the boat. John liked the idea, and discussions began.

I bought a set of Swaggie plans and scrutinized them for hours. But I wanted some things different, and I fiddled with the plans trying to create my own personal boat within John's original hull lines. I found it difficult to get a navigation station, wet locker, galley sink, berths, and stowage shoehorned into 18 feet. So I asked John if I could stretch Swaggie by 12 percent and make my voyager 20 feet long. As the emails flew, however, John asked me what I planned to do once the boat was finished.

"Sail her around the world," I said.

He answered, "I think you really need something different. Something more traditional. Something a little larger. Let me see what I can do."

I agreed.

Two months later, a study sheet arrived in the mail. It showed a stubby, heavy-looking gaff cutter of 6.5 meters LOD.

"That's it," I emailed John. "Don't change a thing."

"I'm leaving Japan for New Zealand," I told the girl. "I'll build a boat there. The children are grown and married now.

I've written more ads and articles and annual reports than I can count. But I haven't sailed around the world, and it's time I did."

"I knew this would happen someday," she said. "Take care, and make sure you take a satellite phone."

Although its form changed, the dream moved slightly closer to reality. Now I had a plan, a sketch of a small yacht that John said would take me around the world, a commitment to rent space in which to build it, and acquiescence from the girl. I started sending money to New Zealand. And, since air fares out of Japan plummet after the summer holiday season, I bought a round-trip ticket with the outgoing leg on August 30th, 2005, leaving the inbound one open, then went to the New Zealand Embassy to get a visitor's visa. I planned to build the boat in a year, then sail off toward Cape Horn in the autumn of 2006. Like I said, a plan. Harking back to my dream from Hawaii, her name of course was *Resolution*.

The future belongs to those who believe in the beauty of their dreams.
--Eleanor Roosevelt

Swaggie

Australian slang for a tramp or itinerant who carries his bedroll, or "swag," on his back.

Here's how John says Swaggie came into being. "My client loves small craft and has long had an ambition to cruise a very small boat capable of blue-water voyaging from his home on the Southern Coast of Australia. For those not familiar with the area, that's roaring 40s territory and there are very long stretches of coast without shelter or refuge. In a storm, the best option is to get as far out to sea as possible, close the hatch, and get into your bunk, but of course few very small cruisers are designed to survive this sort of treatment.

We corresponded about ideas for some time, and we seemed to have similar ideas if slightly different approaches, so I drew a study proposal and sent it off to see what he thought.

Bingo, a check arrived by return mail! Hit the jackpot and rang the bell!

So here is Swaggie!

Swaggie is designed to be sailed from inside. Her junk sail can be hoisted, reefed, sheeted, and struck from the main hatch. As a consequence, she doesn't need the sail-handling areas of conventional craft, and can do without a cockpit, which allows a spacious, comfortable cabin that's much larger than you'd expect in a boat only 18 feet long.

She has a double bunk forward that's much more spacious than you would imagine. Big lockers under the double hold a bank of batteries, 25 gallons of water, and plenty of space for stores and clothing. Sitting headroom at the end of the double makes working in the galley a cinch. There are small lockers on both sides, a galley bench to port, and a general bench starboard with storage beneath both.

Aft of the benches, at the lowest point in the cabin, armchairs on each side offer comfortable places to sit when off watch. John is a firm believer in comfort aboard, and these chairs are as good as it gets, he says. They're handy to the bookshelf and the galley stove, near the on-watch person, but separate enough to allow a nap when you're not on the helm.

Right aft of the armchairs are quarter berths that double as helmsman seats, as they are high enough to bring the eye to window level and offer a 360 degree view of the ocean and Swaggie's sail through a polycarbonate "astrodome" in the main hatch. The helmsman sits in full control of the vessel yet completely out of the weather. Beneath the quarter berths is room for more fresh water; the boat holds a total of 180 liters, which will see two people through a 30-day voyage.

Swaggie's got a portable head stowed behind the companionway stairs, and curtains can be installed to divide the cabin space if a modest crewmember wishes to take a sponge bath in privacy. John has also planned enough space under the after deck for a valise packed inflatable life raft, which is compulsory in New Zealand and some other countries if the boat is to be sailed beyond territorial waters.

On deck, a large well in the bow houses the main anchor along with its chain and rode. The cabin top accepts a custom-designed 6'6" dingy that also protects the skylight while at sea, and the flat area aft of the main companionway is large enough for lounging, or steering with the emergency tiller during fine weather. John suggested a beach chair fitted into on-deck cleats would be just right.

A tall stern pulpit increases crew safety, acts as the mainsheet horse, and mounts the wind-vane steering system. The side decks are wide enough for trips forward, but jacklines should be fitted and safety harnesses worn and secured at all times.

Swaggie's hull form draws on John's proven Houdini design with a narrow, flat bottom, steep deadrise chine panels and well-flared topsides. The fine entry ensures an easy motion and the cross-sectional shape is designed to give a gentle roll with a very high ultimate righting moment. Her ballast is 450kg of lead some 550mm below the waterline, and heeled to

90 degrees she will lift something like 60kg with her masthead, which is a huge righting moment for a little boat. Both safe and comfortable, this is perfect for a craft intended for long voyages where it has to ride out foul weather instead of being able to duck into some sheltered harbor.

Plans are available at http://www.jwboatdesigns.co.nz.

Chapter Two

Resolution, the Sundowner

It's not because of noisy cormorants or bellbird calls, perhaps it's the very silence that forces my eyelids to part. Dawn pushes at the darkness, slowly gaining way.

I sit up and push the thick sleeping bag away. Late March 2018 in New Zealand starts the downslide into winter, but daytime is still warm enough to go without a jacket.

Sweeping the cabin curtain aside, I look out over the Weiti River basin. Almost full flood tide. The quiet water mirrors wisps of clouds, white now, but the rising sun will soon turn them pink. Across the river, another motor cruiser lies quietly at its pier. No one lives aboard, though. Cormorants gurgle their distinctive cry as if trying to clear their throats. Three stand on a sandbar, wings outspread so their feathers can dry.

The white topsides of the boat across the river basin gathers light and stands out against the dark mangrove foliage on the bank. The clouds above the boat turn coral pink and I just can't help dreaming once again of a vagabond life that matches itself to the seasons and the wind and the weather. My mind turns back to *Resolution*, the little live-aboard that was to have been my home around the world and gone. I remembered my dream of her and how I had written it down in present tense long before the first kauri plank was screwed down on her hull stringers. Here's the dreamboat I set out to build those many years ago.

I wrote, pretending the boat had already been built. Now that she's gone, I'll keep the same tone, writing as if *Resolution* the Sundowner were still alive.

John calls *Resolution*'s design the Sundowner class. It's part of the same line of boats as Swaggie. Where Swaggie, the diminutive of Swagman of "Waltzing Matilda" fame, means tramp, Sundowner is more specific. Originally, it referred to Australian itinerants who showed up at a farmhouse at sundown looking for a place to spend the night, but were gone before the day's work started next morning. Nowadays, however, Sundowner has taken on a romantic hue, and connotes those who go where and when they please, completely without ties. *Resolution* is a Sundowner, and in a way, so am I.

A small ocean voyager must be tough and relatively heavy. Tough means decks, cabin, cockpit, and hatches of virtually identical strength. *Resolution*'s a composite of wood and epoxy, and high-quality marine plywood is integral to her strength. John designed her with 12mm thick marine ply frames that are also bulkheads. The frames are bolstered with hardwood and serve as bearers for relatively large laminated stringers. The hull's wide for sail-carrying power, but offers a fairly sharp "V" to the waves at the bow. *Resolution*'s two hard chines add stability and help damp the roll. And the big laminated chine logs add to the toughness of the hull structure. *Resolution*'s skin is two layers of New Zealand kauri glued and fastened over the chine logs and stringers. A tougher combination would be hard to find. Furthermore, the decks, cabin sides, and cabin roof are all two layers of marine ply, laminated together. And every panel is screwed and nailed to the stringers or beams that support it. Very strong, yet simple and doable by an amateur boatbuilder like me.

But a hull needs more than strength alone. It must be shaped properly and designed correctly. John based *Resolution*'s shape very loosely on the Itchen Ferries of England. He gave her 1,540 pounds of lead ballast set in a full-length keel. She displaces over 5,200 pounds, heavy for a 21-foot boat. But then, she must carry enough stores and equipment for ocean passages of up to three months. She's fitted with a Farymann-Bukh 7.5 HP diesel engine that runs up to three-and-a-half hours on a gallon of fuel, and her propeller is protected by the full keel. With a draft of just over three feet, she can anchor inside the crowd of cruising boats that draw five feet and more. And with the forefoot of her keel well forward, *Resolution* sits easily against a piling or a wharf to dry out for bottom painting or other maintenance.

With decks, cabin, and hatches built to be virtually indestructible, *Resolution*'s stanchions, deck hardware, and pulpits are all through-bolted and backed by bronze plates. Her cockpit is small, taking up only about one-fourth her LOA. Four people can sit in the cockpit by design, and the Sundowner class includes an interior layout that sleeps four, but *Resolution* was built for two. The cockpit seats cover the quarter berths for half their distance and lockers are built into the sternmost sections. The shallow cockpit has an access hatch to the stern tube area. And the decked over portion just forward of the transom houses a locker for odiferous items like paint, thinner, and kerosene.

Besides its solid construction, the cockpit has compass, engine controls, pad eyes for safety harness tethers, heavy-duty latches on the lockers and stern tube access hatch, and a tough canvas dodger that kept spray from going down the main hatch.

Simplicity, Simplicity, Simplicity

John's philosophy, which matches my own, is to keep it simple. When it comes to his own boats, he says: "If I can't fix it with what I have on board, or continue on without it, that item does not get put in the design."

Resolution will sail the Southern Ocean. She'll spend most of her time around the 40th south parallel, but to round Cape Horn, she'll have to dip down to 56o south and endure the Howling Fifties. So her rig must be strong and simple. Swaggie has a junk rig, *Resolution* demanded to be a cutter.

A thick-walled aluminum flagpole mast centers *Resolution*'s simple gaff cutter rig. The jib's reachable from the foredeck. Forward, an overlapping genoa rides the head stay on self-furling gear. On the forestay, a loose-footed staysail or a little spitfire storm jib can be hanked on as necessary.

The main sail laces to the boom, mast, and gaff. Its leading edge is long, adding power, and it has three sets of reef points to slab-reef the sail down to handkerchief size. If the wind kicks up, a storm trysail and storm jib help her make way in heavy weather.

Both boom and gaff have jaws rather than goosenecks, and they sit on lanolin-soaked leathers wrapped around the mast. Almost every part of the rig's handmade, and it can be repaired under way without special tools.

The mast is guyed with 3/8-inch galvanized iron wire – shrouds, forestay, head stay – spliced in large eyes that fit over the mast hounds with thimbles at the ends to fit the oversized chain plates tensioned with 3/8-inch toggle-jawed turnbuckles.

Simplicity and single-handing go together. And once the sails are set, everything can be done from the cockpit. Changing sails, however, means getting out on deck. Of course, that's motivation to change early, especially if the weather deteriorates. Light air, however, brings another set of sails—a topsail above the gaff and a big genaker designed especially for gentle breezes.

Resolution carries cloth, needles, sewing awl, grommet setter and extra grommets, repair tape, and everything else she needs for sail repair. Likewise, all the standing rigging can be fixed – everything but making a new mast. She's also got an extra main, staysail, and genoa.

Room for Two

I planned to single-hand *Resolution* around the world, but she's built for two. The two berths are almost exactly in the longitudinal center of the boat, the best place for comfort. On voyages, high lee cloths not only keep the sleeping person in the bunk, but also separate the off-watch crewmember from the rest of the cabin (not a problem when single-handing).

The berths also act as settees with room for two on each, so four friends can enjoy an evening of conversation in the golden light of *Resolution*'s kerosene lanterns. A table folds down from the bulkhead for meals or evening drinks, and a kerosene heater stands against the wall. The galley's gimbaled kerosene cooker and removable sink work well for two under way, and 200 liters of fresh water mean two people can enjoy an extensive offshore voyage as well.

Forward, the hatch provides sitting headroom over the marine toilet and a curtain divides the area from the main cabin. Extra sails go in the forepeak, beyond the head.

At night, the navigation station's illuminated by red or white LED lights, and the navigator sits on a special seat I made to fold up against the center galley wall when not in use. It's a tight fit, which is a good thing when plotting a course in a plunging, mucking boat that's making its way through storm. The tabletop lifts for chart stowage, and has lockers beneath. The VHF radio is on the bulkhead along with the sextant, my three hand-held GPS units, the hand compass, and a satellite phone.

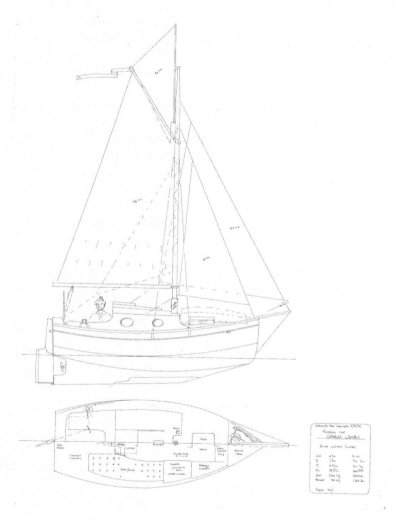

I built the main cabin so it had standing headroom forward of the settees and made a pleasing slant upward towards the main hatch area. No need to walk with hunched shoulders in that spacious cabin.

Minimal Electrics and Mechanicals

Resolution's simple. That means minimum electric and mechanical help. The running rigging operates with multipart purchases, a simple furler system, and two bottom-handle winches. In a pinch, I can crank-start the Farymann-Bukh diesel engine, so it needs no electricity but it does have an electric starter and an alternator to feed the battery bank that powers the few electrical items on *Resolution*'s list of equipment.

Every light aboard *Resolution* is LED for minimal power drain. That includes the tricolor at the masthead, the all-around white steaming light, the navigation table lights, and the compass light. Otherwise, she uses kerosene, for cabin lights, for offshore running lights, and for the anchor light.

The boat also has a 12V accessories outlet that charges phone and laptop and powers a searchlight. The GPS units work on dry cell batteries, with an ample supply of extras sealed in waterproof Tupperware.

Resolution		
Length Overall	6.5 m	21 ft
Beam	2.8 m	9'3"
Draft	0.96 m	3'2"
Sail Area	38 m2	440 ft2
Disp.	2400 kg	5,280 lbs
Ballast	700 kg	1,540 lbs

The Designer's Point of View

As I moved ahead with building *Resolution, the very first Sundowner, I wrote articles for* Small Craft Advisor *in Port Townsend, Washington, USA. For one of those, John put down his thoughts on designing a yacht, Sundowner in particular. With his permission, I include those lucid thoughts here.*

Designing a boat is a real exercise in visualization. Not just in those criteria that make the boat look a certain way, sail well, make her fast, or comfortable or an acceptable combination of each, but the interior and working areas are a visualization in three dimensions plus the imponderables such as ambience and feel.

As *Resolution*, Sundowner hull number one, has her interior built in it's great as the designer to be able to sit on bunks, on seats and in the other spaces to remember what I was trying to achieve when drawing in two dimensions at 1/10 scale, and see if the reality matches the vision.

I tend to be conservative, to add a little extra space here and there, to make sure that the side decks clear the back of the head by more than the minimum, to make bunks slightly wider, slightly longer and with space for a bit more padding than the minimum. This means that my interiors come out feeling spacious, so even in a very small boat a person of average build will find that there is space for the legs to stretch, to lean back in comfort, and to lounge without having to squeeze into the space.

In this case, the boat will be home to her skipper for long periods, and although small will have to feel roomy, has to provide space to sit, lounge, sleep, cook, and work in comfort so the skipper will not accumulate bruises from unplanned collisions with sharp corners, will not feel sore in the joints from being cramped, or stiff from a poor posture when huddled in a seat.

There has to be space for storage, all of which has to be secure when the boat is being thrown about in a violent storm, some of which has to be accessible in any circumstance, and some of which has to accommodate very heavy gear. There needs to be a LOT of storage, as it takes a great deal of food, water, and other consumables for a voyage of this length, and it takes some thought to fit it all in and still retain all the other qualities that are needed inside this little odd-shaped space into which so much has to fit.

I use the diagrams in *Skene's Elements of Yacht Design* for my basics when fitting a boat to a body. They give heights and distances, reach and clearance standing, sitting, or lying down. It's a good set of rules for the ergonomics, for fitting the boat's interior to the human body. I have used them for so long now that I know most of the key measurements by heart, but still sometimes visitors catch me with boxes and bits of scrap plywood, sitting on cushions and leaning, or reaching across a makeshift desk to get the feel of an interior that otherwise exists only on paper or in my fevered imagination.

Experience is a huge help. Having done a few similar boats, it's easy to transfer the experience from one to the other. But Sundowner is special, only 6.5m (about 21ft 6in) long and intended for very long-range cruising. Small additions can make a huge improvement to the feel of the boat, but several small increases can wipe out a useful storage area, or make some other part of the boat too small to be workable.

So it's nice to sit in Charlie's boat as he puts the interior together. To sit inside this massively strong little hull with the comfortable feel of wood, lots of elbow and leg space, the galley close to hand but situated so that a loose pot can't land in the cook's lap and with space for all of the functions of life.

It won't be long now until I can walk and sit on deck and in the cockpit, and start working on the location of winches and cleats, in themselves as critical as the boat's interior. I'm looking forward to it.

--John Welsford,
Designer
http://www.jwboatdesigns.co.nz

John Welsford Bio

As a very small child I began playing with boats in the creek at the back of the house about as far from the sea as it is possible to get in New Zealand, but that's not very far. That continued when the family moved to a new farm at the head of a tidal creek, where, inspired by a story in a "Practical Mechanics" magazine I built the first of many canoes made of roofing iron. My grandfather further fueled my interest by taking me fishing in his 14 footer, then by getting me to help out at a local boatbuilders yard where he worked part time.

With this in my background I was well prepared for the time when I was done with motorcycle racing, was married so the distraction of "girls" had subsided, and I was looking for a sport that would involve both of us.

Sailing was our choice, and that started the ball rolling again.

Building, then racing and cruising our yacht sparked an interest in design theory, and when I was approached by a friend who was showing off a set of plans for a boat that was totally unsuited to the task it was intended for, I offered to design a boat for him. It worked out well and that sent me off on a path that lead to a career in boat design which has seen me lecturing on Marine Design at University, writing for magazines and talking to audiences at boat shows and yacht clubs and travelling the world to work with my customers.

It's a fascinating "trade", filled with interest, applied theory and lots of time out on the water with boats that originated as an idea that I then drew out on paper. The best part of being a "designer" though is the people I meet, and the 2 ½ years when Charles Whipple occupied my workshop while I drew plans just fast enough to stay ahead of his building was one of the best times I've had as a designer, I hope you enjoy his story. I did.

Resolution's *Planned Route around the World*

Resolution was built in Hamilton, in the middle of New Zealand's North Island, about ninety minutes from Auckland, and launched at Tauranga Bridge Marina on the Bay of Plenty. I can't think of a better place to start a world odyssey. First, though, I planned to take a shakedown cruise to Sydney, Australia, sailing across the Tasman Sea and back, and I originally thought to make that cruise in April or May 2006. The best-laid plans . . . . but on with the planned voyage.

Once *Resolution* is stocked with stores for 100 days, along with 200 liters of water, 16 gallons of diesel fuel, kerosene for lighting and heat, three big batteries, and clothing and gear for the voyage, we'll leave the Bay of Plenty and sail the 40th south parallel to about 90 degrees west longitude. There, we'll angle toward 56o south and 68o west to round Cape Horn. Once past the Horn, we'll set a course for Stanley, which is on the east coast of the Falkland Islands, at about 52o south and 58o west.

At Stanley, I hope to take a hot shower and replenish my water and food supplies. *Resolution* will get repairs and repainting if necessary. We probably won't stay very long—perhaps a week, no more than two—before following the old windjammer route from Cape Horn to the Cape of Good Hope at the southern tip of Africa. Landfall will come at Cape Town on the western shore. The leg to Cape Town is only about half as long as the one to Stanley, and we should be able to make it in much less time. Nevertheless, I will take on 200 liters of water in Stanley, and fill all the fuel tanks. And I'll top up *Resolution*'s larders with whatever is available in the island city.

At Cape Town I hope to haul *Resolution* out and store her on land while I make a quick trip to Japan. To keep my permanent residency, I have to enter Japan at least once a year. And depending on the season, I may want to let *Resolution* winter in Cape Town, and we'll sail for Perth, Australia, in the spring.

Restocked and repainted at Cape Town, we'll sail around the Cape of Good Hope and set our course for Perth. From this point, we may go north of Australia though the Timor Sea and the Torres Strait, or we may cut down to the 40th south parallel, sail east past Tasmania, around New Zealand's South Island, and up the eastern coast to the Bay of Plenty. Whichever, my arrival in the Bay of Plenty will mark a successful circumnavigation and may fall on or near my 66th birthday. Once safely back in New Zealand, I'll refurbish *Resolution* and sail her north to Japan, and from there across the Pacific to Hawaii and on to the West Coast of the United States.

The odyssey I planned was a very long journey, but even the longest journey begins with a single step. I got the airline tickets to New Zealand, and that felt like a very big first step.

But why sail around the world alone?

Even a day in an open dinghy is an adventure. The biggest adventure for a sailor is to sail around the world, and the ultimate journey is to sail around alone.

As long as I'm to sail *Resolution* around the world, I suppose there are many reasons to do it alone, but most of all, I wanted to prove that I could . . . even though I was past 60, and even though I was raised in the mountains of Arizona.

I think I can, I think I can . . . .
--The Little Engine that Could

Chapter Three

The Time It Rained

    Back to my first trip: Auckland, New Zealand—August 31st, 2005. My plane landed on time and the airline didn't misplace even a single bag. I shouldered my backpack, which contained a Tamaya sextant, and pulled on the handle of my computer case. It rolled obediently in my wake. My black suitcase with its chartreuse belt arrived at baggage claim, and I whisked myself through the formalities to land without incident in Kiwiland, there to build a boat and sail 'round the world.

    Flying to New Zealand, half a world away from my home in Japan, meant leaving one day and arriving the next, which is confusing at the receiving end. I left Narita on August 30th and arrived in Auckland on August 31st. But when I walked into the main lobby, no one was there to meet me. I finally opened up my laptop, found John Welsford's phone number, called him, and then waited two hours for him to arrive. He assumed I'd arrive the day I left, not the day after

    Everything was new. People in New Zealand talked funny, but looked as polyglot as Americans. Like the civilized folk in Japan and the UK, New Zealanders drove on the left, and roundabouts kept traffic flowing. I figured I could survive, once everything got set up. After all, I was full of hope, in pursuit of my dream.

    All I knew about the country was that their rugby team was the All Blacks and they had a lot of sheep. The drive from Auckland showed me a hodge-podge of green fields, green trees, green ferns, and black tarmac roads. But the only animals I saw were cows; lots of black-and-white Holsteins and a few reddish-brown Jerseys.

Not far from the airport, John insisted on stopping at Pokeno for ice cream. Huge three-scoop cones are the smallest portion available. John ordered rocky road. I got walnut. We were well on our way toward Hamilton before either of us spoke again.

Two hours' drive in John's Toyota Vista put us at his office and workshop in the Matangi area of Hamilton. He ushered me into the unfinished structure. The smooth cement floor seemed absolutely level. And the great open space in the center longed for the keening of saw blades, the whirr of power drills, and the clean scent of freshly cut wood. It made me impatient to get busy building *Resolution*.

No sooner did we get to Hamilton than John hit me with a double whammy. "I'm going to the States in a couple of weeks," he said. "Be gone for 20 days or so. But before I go, we'll get enough work lined out to keep you busy until I get back." I was already scared that my boat-building skills would not pass muster under John's experienced eye, and now he tells me he's not going to be here.

Three weeks with no designer! And not a single sheet of the design for my Sundowner yet drawn. I gave John a weak grin. "Fine," I said. All I had to fill those weeks was a large expanse of concrete surrounded by a bandsaw, a planer, a joiner, and a table saw, along with a wall full of hand tools, drills, sanders, routers, and a bench fitted with vises on each end. The slim Japanese carpenter's saw I'd brought along was a puny addition to John's array of implements, but he'd said the shop was equipped with everything necessary to build my boat. Everything but the plans, it seemed.

When we talked about me going to New Zealand to build the boat John was to design for me, he mentioned that I'd be welcome to stay at his "club." The price seemed right so I jumped at the offer. The Waikato Outdoor Society (WOS) occupies a beautiful green site on the south edge of Hamilton. Besides the clubhouse, five house trailers sat in a line with two cabins on the west side. The first of the house trailers, which New Zealanders call caravans, was for paying visitors, and that's where I lived until I could purchase a home of my own. Here's the problem: I'd come to New Zealand to build a boat, but the first line on the first sheet of the plans was not yet drawn. In the end, the plans for the Sundowner, the boat I built, had something like 40 sheets. True, I came to New Zealand to build a boat, but first I had to get situated.

I arrived with the clothes on my back and a few more in my suitcase, so there was a lot to purchase to make NZ livable. Yes, for the moment, the rental caravan would do. At least for the 20 days John was to be abroad, it would do.

For the two weeks before John left to go overseas, I had his car to get around in, but he'd soon be gone, so I needed my own wheels, if only to get from WOS to John's workshop every day.

When you're chasing a dream, things seem to happen as if the Fairy Godmother was lining things up. We were driving toward somewhere, I don't actually remember where, but there sat a middle-aged Mitsubishi Mirage station wagon with a FOR SALE sign on her windshield. $1,200, said the sign. I got the old lady for $1,100 cash. Suddenly I had wheels. That tired old Mitsubishi saw me through the entire *Resolution* build and then some. She turned out to be worth much more than the NZ$1,100 I paid for her.

One day John said, "Let's go see Malcolm."

I had no idea who Malcolm was, but I trusted John. We went out, hitched John's little two-wheeled trailer to his Vista, and trundled off to Cambridge to Bart's Timbers. That company glued small boards of lumber into large timbers. We were looking for castoffs, misfits that were structurally sound but had some kind of cosmetic defect that caused them to get tossed onto the reject pile.

Some 2x4s glued into 4x8 timbers were just right for my needs. We took the few I'd use from that pile of rejects and left a case of beer and a bottle of wine in exchange. I had the timbers now, but no idea what to do with them.

Back at the workshop, John said, "OK, time you got to work. Take these plans for the jig, use those laminated beams we got from Malcolm in Cambridge, and make it square and level. You'll build your boat on it." He handed me a sheet of drawings. "That should keep you busy for a day or two."

While I labored to get the jig right, John pulled three 4x8 sheets of fiberboard from a stack of material leaning against the end wall. "When you get through with the jig," he said, "use these to make a worktable on top of it. Frame it with

2x3s, screw the fiberboard on the frame, and paint them with that light blue paint." He pointed at a white 5-gallon bucket.

Four days, going on five, I built the jig and worktable. And John stood at his drawing board, turning out drawings on sheets of velum to keep me busy.

The day I slapped a coat of light blue paint on the worktable, John came into the workshop with a smile on his face. "Want to take a ride? Let's go get some lumber."

We drove through pastoral New Zealand, putting Hamilton to our back and aiming for Tauranga on the Bay of Plenty. Sheep and cattle and horses dotted the land, and houses were screened by tall hedges and shaded by cottonwoods, New Zealand kauri, gum trees, and the occasional paulownia. Some place names, like Cambridge and Morrinsville, have a European ring, but most of the place signposts carried Maori words – Tamahere, Matamata, Waipa, Rotorua, Karapiro, Tirau, Tokoroa. While the scenery looked familiar with board and wire stock fences and farmhouses with white siding and stone facades, the names recalled a time when the indigenous Maori knew New Zealand as Aotearoa.

We towed John's little two-wheeled trailer more than a hundred kilometers before pulling into the Moxon Lumber Company at Maunganui. John worked in the lumber and forest products industries for decades, so I left it to his seasoned eye to choose planks from the tiers of air-dried lumber. We came away with a dozen—Fijian kauri, sapele, and tigerwood, a first for me. John said tigerwood is *Hopea Gabrifolia*, from an area that covers Timor, Indonesia, Malaysia, and the Philippines. A dark brown wood with vertical striping, it weighs some 820 kg per cubic meter, compared to 770 kg per cubic meter for white oak. Slippery, wear resistant, and not easily split. Tigerwood's the slickest stuff this side of wet soap with a side order of Vaseline. Coming back from the lumberyard, the damn tigerwood kept slipping through our best rope work. Both John and I have been around boats and marlinspike knottery for decades, but neither of us could make that wood stay put. We did manage to get it to the workshop in Matangi without losing any. But two planks ground off their corners dragging on New Zealand's pea-gravel-and-asphalt paving. Let me put it this way: Hatches sliding on tigerwood won't get stuck.

John left me with the offsets and specifications for deck beams and stem, detail drawings for main and forward hatches, the main gaff and boom, the mast, and the gaff jaws. "That should be enough work to keep you busy while I'm gone," he said, and flew away to a boat raid on Lake Mead, Nevada, a couple of trips to Disneyland in California, and a day in Tijuana, Mexico. Alone, I laminated beams and studied the hatch drawings. Outside, it rained.

I told you somewhat about WOS, the club, right? John casually wrote me just before I left for New Zealand. "Oh, by the way," he said. "The club is a naturalist one. People join to have a place to shed their clothes."

I told him I didn't mind if others went without clothes as long as they didn't expect me to do the same.

"Only in the showers, the pool, and the Jacuzzi," he wrote.

After I moved into the rental caravan, I was the sixth full-time club resident. John, wife, and son lived in the third caravan down, Bruce lived in the second caravan, and PJ lived in a bus. The population more than doubled on weekends with Doctor Wal, wife Val, and Val's sister Noelene staying at the western cabin, Jen and Jeff in the fourth caravan, Bill and Gayle in their camper, while others came and went with the seasons. Notice all the first names. At WOS, first names were all a person needed.

Summer brought Wednesday evening barbecues with sausages and lamb chops on the barbie and everybody bringing salads and veggies and desserts. The car park, as they called it, started to fill up about 4:30, when happy hour began. The cook then fired up the barbecue and laid out the sausages and whatever fresh meat he had (only men barbecued, for some reason). Dress for the evening was work clothes for me and club uniform for the others. Club uniform at the Waikato Outdoor Society would have been called birthday suits in Arizona. The men cooking at the barbecue took care to stay at least arm's length from the coals. Safety first.

Two days after John left, I walked into the clubhouse just before sundown. Gayle and Christine (people go by first name only at the club) were sitting at the main table with Bill.

"Would you like tea?" Gayle asked.

"That would be great," I said, expecting a tall frosty glass.

Gayle grabbed a plate and headed for the barbecue. "Wait, wait," I said. "I already ate. Tea's something to drink, right?"

"Oh, no," she said. "In New Zealand, tea is the evening meal." New Zealand and the United States—two countries separated by a common language.

Still, the meals were excellent and the company was good. I really enjoyed Wednesday evening barbecues at the club. And it was club members who congregated at Tauranga Bridge Marina to see me off when I sailed away.

In the shop, though, I was still on my own

Although I had a sheet full of curves and numbers from which I could laminate beams, the job required much more. Cutting 2x12 planks into 50mm x 8mm strips, for instance. It took six to eight of those strips to make one beam, so I had to rip 80 or so before I could even start gluing things together. And even with the strips all sliced, I still couldn't start laminating. First, I had to fire up the band saw and cut out 15 or 20 L-blocks. I called them L-blocks because they had a long tongue sticking out and they looked a bit like the letter L.

Those L-shaped blocks held the secret of laminated beams and stems. The glue-saturated strakes go on the bottom of the L, which is covered with kitchen wrap, and are winched up tight against the side of the L with C clamps until the glue oozes out.

But first, the grid—scribe a baseline and then vertical lines from the base at set distances. In my case, most of the grids were on 200mm centers, which meant I drew vertical lines from the baseline every 200mm. Then, from John's offsets, I took a measurement for each vertical line, and marked it off on the proper grid line. I'd glued together two 4mmx4mm battens, each about 2.5m long. These served as my spline, which I used to mark the curve made by the offsets. Then came the L-blocks, set precisely so the intersection of the vertical and horizontal L lay on the curve line, facing away from the baseline. Two screws through the bottoms of the Ls held the blocks to the worktable, and a length of aluminum laid in the inside corners of the L told me the curve was fair. Once the jig was set up, I painted the strakes with Aerolite 308 (melamine urea formaldehyde) on one face and an acid hardening agent on the opposite face, laid them on their sides on the L blocks, and clamped them tightly, starting with the middle block and working outwards one block at a time.

The deck beams, with their gentle curves, got laid out, drawn, blocks set, strakes glued, and clamps applied in a few hours. But they remained clamped overnight so the glue could set. That meant one deck beam per setup per day. That doesn't mean I sat around idle, watching the glue set. There were other laminations to do. Gluing 2mm veneers into a curve for one half of the gaff jaws, for example. Again, as I could only glue three veneers to the jig at once, it took me a week to put 15 laminations together. What's more, five thumbs on each hand limited my work speed.

While John was gone, I showed up at the shop every day at 8 a.m., and set to work on the beam and lamination project of the day. Once the beams were in their jigs and there was nothing more to do but wait for the glue to cure, I started ripping Oregon pine timbers to get blanks for my gaff and my main boom.

While I laminated components in John's workshop, it rained. And rained some more. The driveway took a good-sized dip where it entered John's property. By mid-September, the dip was a pond 20m long and five wide. A family of ducks moved in, and got quite upset when I drove the $1,100 Mitsubishi through the pond morning and night. I began to see how so many homes in New Zealand can depend on rainwater collected from their eaves for household use. It comes down by the bucketful.

Normally, rain doesn't bother me. Japan, after all, is a rainy country . . . at least I thought so before moving to New Zealand. In Arizona, where I was born, annual rainfall totals about eight inches a year, and rain is a blessing. In New Zealand, after two weeks of constant rain, I began to think of it as a blight.

A major reason for my hate-hate relationship with rain was the Mitsubishi. It leaked. When I opened the door in the morning, the footwells had two inches of water in them. I bought a giant sponge to soak up the water. After an unsuccessful quest for the headwaters of the leak, I finally drilled a hole in the bottom of each footwell. At least then the water had a way out.

As it rained, the number of deck beams leaning against the wall grew. The stem, the biggest piece of lamination in the project, got three laminations and scarf joints per day. The gaff came together from the cuttings left after the blanks for the boom were in hand. In the end, the gaff was four strips of Oregon pine, epoxied together, planed, and sanded.

The planks of Oregon pine for spars sat out in the rain. They'd been salvaged from a building and were full of nails and even had bolts in some places. I pulled the nails one day when the rain had backed off to a misty drizzle. Then a day with no rain came along.

I pulled out John's huge 12-inch portable circular saw and set to ripping Oregon pine into widths that I could plane down to 40mmx40mm and 40mmx80mm. Once I got the wood ripped, it went inside the shop to dry a bit. Still, the ripping took me the better part of a day, and it would take a couple of hours to plane the stock down to size. Boat-building is not for someone in a hurry.

On October 2nd, I took inventory.

Eleven deck beams stood against the back wall, along with the completed stem. The covers for main and forward hatches, laminated of three layers of 3mm plywood, lay by their jig. Hatch components formed two piles, one for the main hatch, one for the forward. The rounded and sanded gaff awaited its jaws, and the mainsail boom sat waiting for the epoxy on its end to harden.

I'd gotten quite a bit done, but not nearly all that John had drawn plans for.

Leaving the club at 7:00 in the morning, I often turned the Mitsubishi's wheels toward Cambridge. The little town was 16 km away but closer in fact than downtown Hamilton. And in Cambridge, I found Fran's Café.

Within five minutes I knew Gloria's name and Dawn's. All the customers knew them and called them by name. The second time I had breakfast at the café, I met Fran, everybody's mother. Everything on the menu at Fran's is good, like going for Sunday dinner at a favorite aunt's. In the sweet case, my downfall—carrot cake of a special Fran's-only recipe so full of goodness and calories it took both hands to lift a slice from the case. After eating a piece of that cake with a cup of fragrant chamomile tea, I felt like heaven was just next door.

Naturally my pilgrimage back to New Zealand 10 years after the fact had to include a visit to Fran's. The only problem was, Fran and Mik Borsos no longer owned the café. They now catered for funerals from their home in Cambridge. "Come on over any time," said Fran's bright voice on the phone. We made an appointment for two days hence.

Back in the day, all the locals ate at Fran's. Cool showcases displayed handmade sandwiches and a devilish array of sweets (coffee cake, tortes, brownies, layers of peach, custard, and cake, huge cookies, all criminally tempting). The hot showcases held shepherd's pies, several kinds of quiche, a stack of wraps—hot delicacies that kept people coming back to Fran's Cafe.

Art adorned the walls. Paintings by local artists. Carvings. Stained glass objects. A collection of china teapots, some shaped like houses, one like a china doll, faces of Fu Manchu.

In one corner, rag dolls sold for $20 each. An ornate wooden cabinet stood against the wall with rows of hand-filled bottles on its shelves. "Jams and chutneys" the sign said in felt-pen letters, "by Grannie Dunn."

The café opened at 6 in the morning and closed at 5 in the afternoon. Breakfast all day, the sign said. I went for eggs Benedict with bacon in the morning, along with a pot of lemongrass tea. When I got there, Annette, the lady from the bookshop three doors down, was often just finishing her breakfast. Gordon the postman would show up, greeting Gloria and Dawn and Annette. He might be followed by a painting contractor and his crew, five big men who eat all-day breakfasts of sausages, fried tomatoes, hash browns, bacon, two eggs—fried, or poached, or scrambled—and four slices of toast. They'd leave only a few crumbs on their plates and would refill their coffee cups at least three times.

One couple sat at a near table. She with her graying hair pinned back with a barrette. He with his white hair thinning on top. They ate identical tortes, topped with whipped cream and a berry. They leaned forward to speak. I couldn't hear what they said, but they felt like they'd been together half a century.

One day, four young Amazons, each standing five-seven or so, sat around an older man who seems to be a coach. They dressed in spandex and pullover sweatshirts, hair in ponytails, faces clear of makeup, freshly washed, soft and shiny. White teeth flashed and laughter greeted coach's pronouncements. Three blondes, one redhead—coach's graying, balding, hair was cut very short, almost shaved.

Mother and daughter sauntered in. A light beer commercial in Japan featured a svelte model who chugs the beer, then pinches nearly non-existent body fat at her waistline and says to the camera, "A little dangerous." Mom and daughter were more than a little dangerous. Both clad in knits. Both showed rolls over beltlines. Neither was fat in the modern sense of the term. Daughter walked with a wanton sway of the hips. Both looked good.

The place hummed. Gloria with the tattoo on her coccyx and Dawn with her blonde hair held up with a barrette strode in and out of the kitchen, bearing their customers' orders, happy smiles on their faces. Is it any wonder that when I ate breakfast out, I ate at Fran's?

The round tables at Fran's were often filled with gaggles of matrons enjoying afternoon tea and a sweet or two. Laughter was free at Fran's, and the women behind the counter wore smiles—they *liked* you.

At Fran's, I was just one of the crowd. No one commented on my accent. No one dug in to find out why I was in New Zealand. It was just like home.

When time came to visit Fran and Mik at home, John found it with little trouble. He only had to refer to Google Earth twice.

The house can only be described as rambling. It spreads over its suburban plot, partly hidden by darkly stained board fences, offset by brilliant blossoms. Roses, Fran said, and petunias, begonias, violas, pansies, alyssum, and a variety of daisies.

The true beauty of the visit, however, came in the sitting room. Fran had baked a famous Fran's Carrot Cake just for my visit. I was overwhelmed. I must have licked my lips because no more had John and I seated ourselves than we found inch-thick slices of carrot cake topped with cream cheese icing sitting on ceramic plates before us, begging to be consumed. Little conversation ensued as we took satisfying mouthfuls of Fran's Carrot Cake. With that cake, for all intents and purposes, my pilgrimage to New Zealand was a tremendous success. And my tongue is only partially in my cheek.

When he departed abroad, John left a sign on the office wall. In huge blue letters, it read COMING HOME Oct. 3, 6 AM. The message was clear. Be there.

I set the alarm for 3:30 a.m., ample time for the two-hour drive to Auckland International Airport. By the time I got to the arrivals lobby, the 6 o'clock flight from Los Angeles had already landed and processing had started.

At 8 o'clock, the passengers on the flight had cleared immigration and customs and left the terminal. No John.

His message definitely said 6 a.m. I tried to check passenger lists but the airlines refused to give out the names of the people on their flights. Another flight was due in from Los

Angeles at 8:45, but surely he was not on that. Perhaps he'd gotten the date wrong. At 8:36, I left the arrivals area and drove two hours back to the shop in Matangi. I figured he'd call the workshop when he arrived.

He drove up in his wife's RAV4 just after noon. He'd landed at 8:45, taken an hour to get through formalities, and then exited the arrivals gate to find no one there to meet him. He'd negotiated a good price for a taxi ride to Waikato for him, his wife, and his son. He waved off my apologies. "Not to worry," he said. "So, how you coming along?"

I led the way into the shop. Thirteen beams and a laminated stem stood against the back wall along with two hatch roofs. A boom blank hung from the ceiling over the equipment racks and the gaff blank leaned in the corner. The rough laminations for the gaff jaw lay on the freshly painted worktable. "I didn't get the mast made," I said. "I don't have enough good Oregon pine. And the gaff jaws aren't finished. I didn't get everything all done," I confessed.

John grinned. "I just wanted to make sure you had enough to keep busy while I was gone. Looks like you had enough to do."

Those first days I was down to about 95 kg from 105, getting up at 6:30, taking a walk for 45 minutes or so, eating a breakfast of All-Bran and milk, and hitting the workshop before 9:00. I usually worked until 6:00 p.m. or so, later as the days got longer. Unlike Japan, people in New Zealand go to work at 8 or 9 in the morning, and quit at 4 or 5. Even the shopping malls close at 5:30 p.m., except on Friday nights. In Japan, the workday starts at 9 a.m., but it often stretches into the night, with people finally leaving their offices after a dozen hours at their desks.

What did that say about my leisurely build of *Resolution*?

I'd planned a year to build the boat, and I'd squandered the whole first month on a few laminated beams, a couple of laminated roofs, a boom, and a gaff. Somehow it didn't seem like I was going fast enough to finish the boat in my one year. Not only that, but I was in New Zealand legally only until June 30th, 2006.

That said, it was relaxing to be in New Zealand, where the job was not the sum of existence. At the same time, I wasn't getting as much done as I thought I should. I'd been in the country a month already, and my boat didn't have a set of plans. I think I knew, even at that point, that I definitely wasn't going to finish that boat within a year. And in Kiwiland, it just kept raining.

Into each life some rain must fall.
--Henry Wadsworth Longfellow

Chapter Four

Bits and Pieces

Thousands of years ago, Chinese philosophers turned many small steps into journeys of great distances.

It was mid-October and the first of the plan sheets for frames was finished. I took it to the copy shop for a full-sized reproduction—at $4 a shot. Later I tried to save a few bucks by making drawings in a lined tablet I used as a workshop book.

Speaking of plans, John had some interesting things to say about the process of making the plans for *Resolution*, and by default, the entire Sundowner class. He wrote:

"In the *Resolution* project, I was able add to the drawings and written information. I also had the opportunity to sit and chat with the builder; he could walk the few steps from where he was pushing bits of wood around and ask me questions. I sometimes spotted things as I watched him at work and we'd discuss my intentions while talking over the next stage of the build.

"This gave us several channels of communication, the plans themselves being only one. But I had three other customers getting started on their Sundowners, and with one each in Australia and Brazil, and another some hours' drive away, I had to work a bit harder to convey my intentions to them, otherwise all of us would have been endlessly corresponding.

"Most of those who will join the Sundowner family are anything but expert boatbuilders. The boat is designed to be easy to build, it doesn't require big fixed machinery to slice up the big pieces into smaller ones, and my intention is that anyone who can read, measure, cut to a line and use very basic woodworking tools should be able to see a dreamed-of destination come up over the bowsprit of their own little cruiser a couple of years after buying the plans. (Insert your own dream scene here, whether palm trees or snow-capped mountains above a glassy, still fjord!)

"So each bulkhead or frame that I drew had its own full sheet of tracing paper. It was drawn at 1/5 scale so the picture is large enough to have lots of detail without getting too fine in the detail, and on each sheet of plans there are several full-scale drawings of key joints and assemblies.

"To know what others might ask means I must try and determine what it is that they bring to the building: what tools, what skills, what knowledge, and what attitudes. The designer's job when dealing with a professional is easy; I can assume that they know certain things, work in certain ways, and to certain standards. They will have certain tools and other known resources. An amateur, though, is a much more challenging prospect, so in order to make the job achievable, what I must produce is almost a lesson in boatbuilding, and that's what I do—a lesson in building this particular boat.

"Working on the spot with Charlie was great, a lesson for me as well as for him, and everyone else who builds a *Resolution* will benefit from our interactions.

"Yes, it took time, but it saved both the builder and the designer (me) time in the end."

Naturally, I the builder started with Frame No.1, the only one covered in 12mm plywood, top to bottom.

Most multi-chined boas have straight sides except for the forward-most frame, which has a bit of a curve in the lower panel. My *Resolution*, or Sundowner, as John named the design, was not so simple. Every side panel was curved, with only the bottom flat. Curves are pleasing to the eye, but tougher to make. At the top of each frame sheet, John drew the panel curves full size. So when drawing the frame out on the worktable, I first cut the curves from 3mm fiberboard, then used them to draw the frames. Of course, I also used them to draw curves on frame arms and on the plywood paneling and doublers. To look at Frame No.1, you'd never know *Resolution* was a multi-chined boat. The laminated stem piece slots into the floor of the frame, so a single fair curve runs up each side from floor to sheer. And at the top, I got to use the first of the deck beams I'd laminated. I epoxied it to the after face of the plywood bulkhead.

I spent the better part of a day drawing the frame on the worktable, laying out the frame arms (20x100mm Fijian kauri), and marking the bulkhead panels. In fact, I probably averaged three days per frame, including the transom; slightly more than two frames per week, eleven frames in all. At that pace, John could easily keep ahead of me with the drawings.

. The second frame back from the front stem had half a bulkhead. John said he wanted it to be waterproof, a dam, in effect, against any incursion from outside in the front part of the hull. Later, we would put a layer of tough Kevlar on the outside of the hull for further protection. In the end, of course, it didn't matter.

Putting full-size 12mm webbing across the frame to form a bulkhead takes time. Putting doublers in strange and wonderful shapes in the frame is even more time-consuming. Frame No.2, for example, not only had a half-height bulkhead across the lower half, it also had battens placed to take the fore-and-aft plywood that forms the faces of the lockers on each side of the forward cabin.

Frame No.3 was open in the middle down to its 70x120mm hardwood floor, but it had the after ends of the lockers placed on each side. In fact, every frame was much more than merely a set of ribs to accept the chines and stringers. Each frame included part of the interior cabinetry, part of the engine bay, or part of the cockpit.

As I stacked finished frames against the wall of the workshop, I began to see the shape of the boat. Although I was well into my fourth month of boat-building when I finished the transom frame, I could feel the progress. Naively, I thought that once the basic parts were done, building would suddenly zoom ahead. It never did.

In spring, New Zealand is a flowered land. The big paulownia trees with their long strings of lavender flowers stood out to me, and the hydrangea, which are also common in Japan, bloom with a vengeance in Kiwiland. Fruit trees flower everywhere—apples, plums, cherries, peaches, *nashi* Japanese pear-apples . . . you'd think the entire country was a huge orchard. Perhaps it is.

The Waikato Outdoor Society sports a huge lawn that looks like it covers about five acres. Often, when I came home in the evening, PJ the truck driver would be on the riding mower, chopping down the week's growth of lawn. The ride-on didn't have a catcher, so PJ created windrows of cut grass as he sped around the grounds. Bruce and I and whoever else was around got the tiny garden tractor, hooked a little trailer to it, and drove it up and down the windrows, scooping up the grass clippings. There was easily enough grass to feed a good-sized horse, but if a horse mowed it, we'd have to shovel manure. Sometimes, you just can't win.

Speaking of Bruce and PJ, they're the mighty hunters of the Waikato Outdoor Society. Long, long ago, white settlers in New Zealand brought brush-tailed possums from Australia with the idea of starting a fur industry. Some of the varmints escaped, and now an estimated 60 million wreak havoc among native New Zealand trees and wildlife. In short, possums are pests. So Bruce and PJ go out at night with a flashlight and a .22 rifle to get rid of the ones audacious enough to venture onto club grounds. But they still eat lemons off the tree by the clubhouse and apples from the crab apple trees down by the lower hedge. No matter how many critters the two hunters bag, there are always more.

Those possums, by the way, sport glorious pelts with hollow hair that insulates wonderfully. Socks and watch caps made of possum hair keep out the cold of southern New Zealand's winter nights like nothing else. The fur industry thrives, but possums in the wild thrive faster, and no eradication program seems to work. Possums are to New Zealand as rabbits are to Australia—harmless-looking, cuddly little animals that cause great damage.

All in all, the first frame stood on the strongback jig nearly three months after I'd taken a deep breath and started set to work. In late November, the day came when I ripped the pale blue worktable apart and readied the jig to accept the upside-down frames. I know, that sounds like a simple job, but the forward two frames and the stem had to be located on the floor with 2x6s drilled and bolted to the concrete. Not only that, but the entire structure had to be straight and true and vertical and horizontal—almost more than an amateur boatbuilder like me could do. And I thought the advertising business in Japan was stressful . . . .

Then I found that frame No.3 and No.4 were built wrong. Fortunately, there's a Welsford theorem that covers those mistakes—actually, two of them. The third Welsford theorem of boat-building says: "You won't find the mistake until after the glue has set." The fourth Welsford theorem of boat-building says: "The mistake that cannot be remedied with epoxy and wood has yet to be discovered." I set about remedying the mistakes.

No.3 was easy. I just had to epoxy a curved packer on the edge of each upper arm to bring the sheer point outboard by 70mm.

No.4 was a bit more complicated. The remedy entailed cutting an arm off each side of the frame. I bisected the middle chine angle with a saw, tilted the arms outboard until they reached the proper sheer point, then put wedges inside the now triangular cuts at the middle chines and backed them with doublers made of 12mm plywood. Frame No.4 turned out a third thicker at the middle chine than most of the other frames, but wood and epoxy and my trusty Japanese saw took care of the mistakes.

The day I got all the frames standing on the jig, John Welsford came wandering into the shop, as he had a habit of doing at most inopportune times. In my mind, I was ready to start notching the frames for the chines. John soon set me straight on that.

"Nice job," he said, putting his hand on frame No.5. "Now that they're all standing upright, you'll need to brace this one so it can't move and then measure to every other frame from it. No.5 is your touchstone. Don't measure to No. 6 and from there to No.7. Measure all of them from No.5. Port, starboard, top and bottom. Make sure all the numbers are right before you go on to the next frame. And when a frame measures right, brace it well, so it can't move. Got that?"

I took my nine hundredth deep breath and nodded.

First I made sure No.5 was right. A spirit level showed that the frame stood straight up and down and sat level on the jig. Then I used a plumb bob to make sure the frame was centered. A string to the forward center point of the jig measured the same distance to both port and starboard gunwale points. After all the tiny adjustments, I braced the frame in four places in addition to the legs that were bolted to the jig. Only then could I use No.5 as the touchstone for all the other frames.

Four days after I'd decided to start notching for the chines, the frames were measured, trued, and braced. Finally ready for chines, gunwales ('wales), and stringers. I scarfed two long pieces of castoff fir together to make a batten that would reach from stem to stern, then bent it over the portside chine knuckles. That showed me where to cut the notches. Let me tell you. I'm now as good as a professional notch cutter. I've cut notches for two chines, four stringers, and two 'wales on each frame from bow to stern – that's 178 notches. Not only that, but each notch had to be done twice: the first cut marked from the batten and another cut to fair up the chine log or stringer laid into the notches.

Now you'd think that 16 long pieces of wood comprise the stringers and such. Not in John Welsford's boats. Chine logs, which are often just called chines, are three 20x25 mm strakes, laminated to make them long enough go from the notch in the transom frame to the rabbet on the stem. In other words, I laid 32 strakes to get 16 chines, stringers, and 'wales. As the year 2005 got ready to turn, I threaded the first layer of chine around the boat. But when I finished, *Resolution's* skeleton stood solid and ready for its cold-molded planking— two layers of 8mm Fijian kauri. Just in time for Christmas.

Christmas in New Zealand is a big holiday. "The Christmas Song" may say "...and folks dressed up like Eskimos," but Down Under, Christmas is bikini time, and the holiday is spent at the beach. And when Santa makes an appearance at Centre Place plaza, he sweats under red cap and white beard. For me, Christmas was just like any other day. But December 26th was a bit special, as my first grandson was born that day. He was soon nicknamed Prince and grandmother and mother leaped to fulfill his every desire. I went on building the boat.

When I moved to New Zealand, I brought along my loner's mentality. I'd lived in Japan since 1977 and tried several times to mix with both local Japanese and expatriate American societies. Didn't work. Much of the fault was undoubtedly mine, but several incidents left me with feelings that my fellow humans were basically an untrustworthy bunch, and I would be far better off by maintaining a distance from them. Of course there were a few exceptions, but they were few indeed. So instead of spending my evenings socializing and getting to know New Zealanders, I prepared a simple meal for one in my rice cooker, ate it with a glass of water on the side, and spent until 11 p.m. or so reading by lamplight in my caravan. Needless to say, my favorite shop in all Hamilton was Whitcoul's bookshop where paperbacks on the special table were six for $20. When I finished a paperback, it went into the club's library.

I rose at 6:30, ate a breakfast of All-Bran, banana, and milk, and took a morning walk of about 5km. I didn't ignore the other people at the club, but didn't go out of my way to socialize with them either.

In his autobiography, James Michener wrote: ". . . as I watched the actors and writers who took advantage of the free life overseas, I was distressed by its effect upon their work. . . . Writers tended to suffer in their own way. They lost touch with America and American themes. They wrote lavishly, or under pressure from distant agents, or in areas they would not have touched had they been back home under the eyes of publisher and counselor. In the worst instance, they became rootless expatriates, yearning for home but afraid to go back . . . ."

I saw myself mirrored in that paragraph: Three decades in Japan writing lots of words but few I'd want to show anyone. None of the writers who did studies on Japan ever walked the world stage. Nor did I.

The *Resolution* project gave me a way out, but once in New Zealand, I was still my reclusive self.

One of the many folks who came over to the shop to investigate the stupid Yank who was making a little 6.5-meter boat for a circumnavigation was John Leathwick, a marine biologist.

Leathwick spent a year building a Welsford design called Navigator, a little 14-foot 9-inch sailing dinghy, and one fine summer day in January, he invited me and John Welsford to come along to Hamilton Lake for a sail.

Mid-morning on a bright January day with a light scattering of clouds and a brisk breeze, John and I climbed into his Vista and drove to the Leathwick residence. John Leathwick's spic-

and-span Navigator *Hautai* nestled on a trailer in his garage. With one hand on the trailer tongue, John rolled the boat outside and hooked it to his family sedan. We parked the Vista on the lawn and scrambled into John's car for the short drive to the lake.

Hamilton Lake, or Lake Rotoroa in Maori, covers 54 hectares and averages 2.4m deep. Still, it has an active yacht club, and a concrete ramp for launching boats.

John's pristine *Hautai* rests on its own beach trailer atop the galvanized trailer hitched to the car. The fat tires of the beach trailer make it easy to push over sand or rough concrete, and after 30 minutes rigging the boat and taking a reef in the main, we gently pushed her into the lake and got ready for a stiff sail.

The Navigator is a flat-bottomed lapstrake design that looks brilliant in the water. John Leathwick's boat used a yawl rig, so it was propelled across the lake by three taut sails. He and I took first outing. We tacked north across the lake and south back. With some 20 knots of wind, *Hautai* virtually surfed the surface of the lake with a two-foot roostertail coming up in her wake. They say you can tell a happy motorcyclist by the bugs on his teeth. If the mosquitoes had been out that day, they'd have been plastered all over my incisors.

We tacked back and forth for some time, showed a laser how to sail fast, and generally collected nice comments from the people who gathered along the shore to watch the prim little Navigator do her stuff. Too soon, I had to get off and let John Welsford take my place. After all, he did design the boat, and she was showcasing the best of traditional sailing that blustery day on Hamilton Lake.

Day after day, I labored to build the bits and pieces that would morph into my blue-water boat. There are no days off. (Well, I went to the classic boat regatta at Mahurangi, where 800-plus wooden boats gathered, some of them gorgeous classics. What's more, 14 of John's boats lay on the beach within sight.) With all the chine logs, stringers, fender wales, and gunwales attached to the frames, all that was left was to give *Resolution* two layers of 8mm kauri planking. The first of the planking got fitted to the topsides five months after I walked into John Welsford's workshop with the idea that I would build a boat and sail the world.

Each day I dry-fit planking, backed out screws, epoxied chines, stringers, and 'wales, and screwed the planking back in place, but I'll talk more about that in the next chapter.

The time will also come when she must be fitted with a lead keel weighing 700 kg. But until then, building *Resolution* is just a matter of cutting and fitting a million or so bits and pieces.

Aim for the sky, but move slowly, enjoying every step of the way. It is all those little steps that make the journey complete.
--Chanda Kochner

Chapter Five

Planking for the Hull of It

I started building the hull of my *Resolution* on October 12th, 2005, when I drew out bulkhead No.5 full size on the pale blue worktable in John Welsford's workshop in Hamilton. From then on, I did little other than work on my boat.

People were amazed that I planned to sail around the world in a tiny 21-foot yacht, but they didn't realize that John Welsford is no novice at designing blue-water boats. Many of those who know John know him for the *Navigator*, the sailing dinghy designed for amateur boatbuilders, for which he's sold more than 700 sets of plans. But the people who keep tabs on the ocean racing scene know that John designed a Mini-Transat boat, also 21 feet, which finished third overall in a race across the Atlantic, with Chris Sayer at its helm single-handed. At the time I was building *Resolution*, John was the seventh winningest designer of Mini-Transat boats on record. So yes, he knows a bit about ocean-going yachts.

The frames of *Resolution*'s hull started going up on the strongback jig on November 29, 48 days (subtract a week for when the girl came to visit) after I started making frames. That averages out to about four days per frame. While it seems a simple task to bolt finished frames to legs that attach to the strongback, the job took me until December 16th to complete.

"The boat's frames are solid as a rock," says my journal for the evening of the 16th. To keep the frames in place, I had battens tacked all around the outside of the frames and braces inside them. I thought the structure was solid then, but I had no idea. "At this rate, planking will start before the year is out," I scribbled. Ha ha. Little did I know. After notching every frame eight times on each side and epoxying 16 strips of kauri on each side from stem to stern, I found that the calendar read January 23rd, 2006. But let me tell you this. I thought the hull felt solid when I had it braced with battens. With all the chines and stringers in, I could climb all over the structure without it moving even slightly out of true.

A single-hander's boat is all he has between himself and the deep blue sea. And a boat is only as good as its hull. *Resolution* is no ordinary 21-foot gaff cutter. She's designed to exploration-grade specs. The first glimpse of this heavy construction came when the chines and stringers were all in place. The smallest brace measured 25x45 mm and the chines were laminated of three 25x20 mm strakes to make them 25x60 mm. Tough. Tough. Tough.

The first plank generated a celebration. It slanted aft from sheer to lower chine and covered 200mm of stringers and chine logs. Out came the cameras and the internet celebrated with us. John put a piece on his site and I wrote in my blog. How big a celebration is that? Worldwide party, that's how big.

John and I cut the planks at Malcolm's place—Bart's Laminated Timber. Malcolm let us use his huge bandsaw for a couple of hours, and we ran 20 2x8s through the saw, ripping off 10mmx200mm strips. Each 2x8 yielded five strips, which after going through John's Chinese thickness planer became 8mm planks for the hull.

Resolution looked like a great big birdcage, but with the stringers and chines glued and screwed to the frames, transom, and stem, the structure was solid. If it were planked, we could have turned the boat right side up and floated it down the Waikato River.

Actually, the city of Hamilton, New Zealand's fourth largest—large is relative here; the population of the entire country could fall into Yokohama and never be seen again—owes its founding to the Waikato River. Back when roads were just wagon ruts through the bush, steamboats carried people and goods up the Waikato River. The docks at the end of navigable Waikato were in what is downtown Hamilton today. Goods and people still gather in Hamilton before spreading out across the Waikato region to towns like Te Awamutu, Matamata, Otorohanga, and Ngaruawahia. Of the 22 towns in the Waikato region, 15 have Maori names. About 20 percent of the population in the Waikato Region is of Maori ethnicity. In ancient times, Maori built huge double-hulled voyaging canoes. In the spring of 2006, less than 50 miles from the home of the Maori king, I began planking my small voyaging monohull. As with every step in building *Resolution*, planking the hull turned out to be a slow, laborious process. The planks lie at an angle to the stringers and chines, and, depending on the curve of the hull, three or four stainless steel screws must go through the plank and into the chine or stringer. Sounds simple, but . . . .

The first plank determines the angle for all those that follow. I clamped it in place, fore and aft, top and bottom. Then drilled pilot holes for the screws. Then drilled countersinks for the screw heads. Then got the cordless drill and a box of 500 3/4-inch 8-gauge screws and screwed one into each pilot hole. The typical plank had four lines of screws, three or four screws to the line. Once I'd sunk a dozen to sixteen screws, I could take the clamps off the plank.

Finished?

Not even close. The edges of the planks were planed off at an approximately 5 degree angle and the widest part of the plank had to be against the chines and stringers. Over three or four hours, I could dry-fit six-to-eight planks. I made sure they fit the chines and stringers correctly, made sure the chine edges were planed off to the right angle, made sure the planks laid tight in place without major gaps between them. Reaching inside with a pencil, I marked lines on the planks where the stringers and chines crossed.

After numbering the planks with a pencil, I backed all the screws out of the dry-fit planks with the cordless drill. Between screwing them in and backing them out, a number of screw heads deformed too much to be reused. By the time I got through planking the hull, the reject box held a few hundred useless screws.

I made up a pot of epoxy and mixed in a judicious amount of microfiber thickener. On the inside of the plank, I slathered thickened epoxy between the penciled lines. On the chines and stringers, I spread epoxy the width of the plank. With thumb and forefinger, I hand screwed a stainless screw into each screw hole. With the drill, I turned the screws in the middle line until they protruded through the plank by about 3mm. This let me position the plank.

Holding the plank by its epoxy-free areas, I laid it on that middle stringer and wiggled it until the protruding screw points found the dry-fit holes. The drill made short work of driving the screws through the plank and into the stringer. Putting my weight against the lower portion of the plank made it lie down against the next stringer while I screwed in the fastenings. Used the same technique on the upper portion, and so on until the plank was back in place as it was dry fit. Repeat the process six or eight times, and the day ends as the last plank is fastened in place. Finally, a swipe with a paper towel folded in half and wrapped around a forefinger removes excess epoxy at the inner joins and leaves them with a slight fillet of the adhesive to keep water at bay.

Whew.

To keep the hull balanced, I'd do one batch of planks on the starboard side, then one batch on the port. In less than two weeks, I had planks running all the way from sheer to lower chine, and that brought on another job that I'd been niggling about in my mine—planking the bottom.

Before I could do that, I had to fair the lower chine to match the angle of the frames, and plane the hardwood floors to match the curve of the bottom of the boat. More time. More patience. More baby steps.

Actually, the bottom received two layers of 9mm plywood, but at first, I needed only one. The first layer required three pieces of plywood, each with serrated edges. The serrations were 30mm wide and 30mm deep, and dovetailed with the sheet next to it. I laid down the first piece of plywood so the serrations landed on the 120mm floor at Frame No.4. With the piece dry-screwed in place, I slipped the next piece under the serrations and penciled their form onto its upper surface. I soon developed a rhythm. Take the piece of plywood off the boat, cut the serrations with a sabre saw, lift the piece back up onto the bottom, and fit the serrations. Put a few screws in to hold the second piece in place. Take the serration form to the end of the piece and mark a set of serrations on the piece so they match the floor of Frame No.9. Cut the serrations.

I repeated the process until all three pieces of plywood were serrated and fit in place. Then went back to the first piece to epoxy it and screw it in place, and continued doing that until all the plywood is glued and screwed securely to the bottom. I wiped off the serrations where epoxy has been squeezed out, and then went inside the boat to wipe each floor where epoxy has also squeezed out during the screw-down process. I also left a little bit of plywood hanging over the edges so it could be planed down to fit the chines at the proper angle.

That's a lot more about planking a boat than anyone but a boatbuilder ever wanted to know, I'm sure. By February 1st, I'd started applying the first layer of 8.5 mm Fijian kauri planks, dry-fitting first, then removing, applying epoxy, and screwing back in place. Twenty-six days later, the first layer of planking, including 9mm plywood bottom planks, was complete. *Resolution* looked like a boat.

Taking a break from building *Resolution,* I returned to Japan for a couple of weeks to see the new grandson and spend some time with my family.

My Other 21-footer

Resolution is not the first 21-foot boat I've owned. In Japan, my friend Masaya Kinoshita and I bought a Blue Water 21 in March of 2001—a classic that had gone out of production. We got hull #274 for slightly more than a thousand dollars. By summer, we'd scraped the old paint from the hull, applied two coats of navy blue polymer yacht finish, and christened her *Millennium Rhyme.* The last Saturday in September, we fitted a new jib furler system to the forestay, put aluminum rails around her sheer, and put a new outboard motor bracket on the cockpit well. We left her snuggled happily to her anchor buoy and went home for the day.

Typhoon No.9 (no names for tropical hurricanes in Japan) was headed northeast, and the weatherman expected landfall far south of us. Instead, the typhoon veered north and came ashore just south of Yokohama, about 50 kilometers from Sanbanse, where *Rhyme* swung at her buoy.

Winds of up to 60 meters per second whipped around the headland and across Sanbanse. Obediently, *Rhyme* faced into the wind to present her lowest area to the onslaught. She was anchored to a fifty-five-gallon drum full of concrete and scrap iron, and had a scope of about four to one, which means her anchor rode was four times as long as the mean tide depth. The headlands protect Sanbanse from monster waves, but nothing can protect the tidal flats from storm tides.

As the eye of No.9 approached, the water rose. Soon *Rhyme* was at the end of her rode. Still the water rose. *Rhyme* went down by the bow, but her displacement was far greater than a barrel full of cement and scrap iron. With each wave, *Rhyme* hopped the barrel off the bottom and inched her way toward the seawall. The water rose. The hops got longer, the seawall got closer. Then she was banging on the wall. She didn't hit the wall broadside so her fenders could protect her; she rode on waves that broke over the seawall. At last the wall tore her cast-iron keel from its fastenings and ripped open her belly. Without her keel, she lost her balance, rolled into the wall, severed the stays, and scraped the mast from her deck against the concrete. She settled in the mussels at the base of the seawall, smashing her torn flanks into the unforgiving steel sheet piles.

We found her there, visible only at low tide, with a motorboat across her bow. We salvaged the sails and pulled the outboard up, just in case. Little else could be saved.

Millennium Rhyme lay dead, something portending the future, perhaps.

Memories of Sanbanse

While I was in Japan, I went down to the tidal flats called Sanbanse where *Millennium Rhyme* had been moored. When I returned, I wrote my memories of her in my journal.

Today I drove to Sanbanse and walked the seawall for a mile or two. Years ago, I'd clamber down a 2x4 ladder to a makeshift dock and paddle a dilapidated rowboat out to where *Millennium Rhyme* snuggled her buoy, waiting. Her 6-

horsepower Yamaha putted us down the dogleg channel between beacons that doubled as cormorant perches. Bamboo poles marked the shallows, starboard and port. As the outboard muttered, I bent the mainsail on mast and boom, hanked the jib, and led the sheets back to the cockpit. The cormorants watched us pass, then dived at the silver flicks of sardines. Beyond the yellow buoy I raised the main and jib, hauled sheets to stretch sails taut and shape them to drive the boat. I pulled the outboard kill switch and silence overtook us. The breeze hummed in the rigging and wavelets shushed against the hull.

> *bright white sails*
> *blue sky deep blue sea*
> *far horizon*

I look across quiet Sanbanse, no longer home to boats on buoys. A lone fisherman sets his line, hoping to catch goby. Perhaps he dreams of tempura for dinner. From the seawall to the *nori* nets in the shallows a hundred yards away, the water reflects the skyline without a ripple. No clank of halyard against aluminum masts. No whiffle of wind over tight stays. No clonk of oars against rowboat locks. Beacons still mark the dogleg channel, and cormorants still dive.

Layer upon Layer

Resolution's hull took the first plank of its second layer of 8mm kauri on March 19th, 2006. That second layer took even more time. Here's how I explained the process in my blog at the time.

The first step in epoxying the second layer of 8mm Fijian kauri planks is to apply a thin layer of epoxy to the hull. Here's how you do it. Before you remove the dry-fit second layer planks, you pencil their borders on the first layer so you know where to use your plastic spreader to smear the thickened epoxy.

The second step is actually two. On the second layer plank, you first spread a thin coat of epoxy over the entire inner area. Then you lay down snotty-looking globs of epoxy thickened with microfibers. After that, you take a 3-inch (100 mm) serrated scraper (which you can buy at a paint store) and create ridges of epoxy over the entire plank. This works to eliminate voids between the first and second layers.

The plank you epoxied was dry-fitted, so it has all the screw holes. Holding the epoxied plank up with your left hand (fingers spread and balancing the plank like a waiter balances a tray of dishes) use your right hand to start the screws in their holes. Screw the lowest ones through far enough for the points to stick out 3 or 4 mm through the epoxy. Holding the screws, place the plank on the wetted out area on the hull, aligning the protruding screws with the holes in the first layer of planking. Use your cordless drill with a square bit to drive the lowest line of screws into their holes. Climb a couple of steps up your ladder, push the plank down at the top, aligning the edge with the edge of the last plank you fitted, and drive in the top line of screws. Then you can drive all the other screws in sequence. This will force epoxy out any holes or gaps and you'll need to clean the blobs off before applying the next plank.

It worked like a dream. When I cut out a piece later, there were no voids between the layers at all. *Resolution* had the toughest hull of any boat John Welsford has ever designed.

Now the second layer was in place and the bottom had another layer of 9mm plywood as well. The chines are planed and rounded, ready for 200mm-wide biaxial fiberglass tapes to be applied. In the corner stands a roll of yellow Kevlar cloth, ready to cover the hull from stem back to the chain plates, crash pads to protect the boat from the sharp edges of semi-submerged containers gone adrift, growler icebergs, or the odd whale. Then, after fairing with epoxy and microlite fibers, we can cover the entire hull with 10 oz. fiberglass.

Resolution was rock-solid. I'd planed off the edges at the lower chine so the second layer of bottom plywood could overlap to the outside of the kauri planking.

2006 saw John teaching at Massey University's Department of Transportation Design in Auckland. That meant a commute to the big city every week, but John said he enjoyed the students, and I imagine they learned a lot from him.

I watched New Zealand's version of eBay, which is called TradeMe, quite religiously, looking for bargains in hardware and fittings. I'd found a pair of Murray winches for NZ$150 each, a real bargain, and needed to pick them up in Auckland, so I went along for the ride when John drove to Massey U.

We picked up the winches at Boat Books Limited (spent too much time drooling over the books), talked to weather guru Bob McDavitt, and met Jim Lott from Maritime New Zealand, who felt a 6.5-meter yacht like *Resolution* should have no problem with circumnavigating. Of course, in the end, circumnavigation was not her problem at all.

During the couple of hours we spent at Massey helping students with their projects, I got to see instructor John in action. What a patient man. No question was too trivial for a straight, often detailed, answer. No idea from a student ever got put down. Any progress was noted and praised. I could see that John had earned the respect of his class, and rightfully so. How few of the professors I studied under at university and grad school had John's rapport with their students. Although he doesn't have an advanced degree, John would make an excellent university instructor, and Massey would have been better off to hire him as a regular instructor instead of as an adjunct.

"Stop off at Pokeno?" John asked as we left Auckland. The smile on my face was answer enough.

We stopped at the ice cream shop for three-stack cones. John got his normal rocky road, and I chose rum 'n' raisin. As usual, we were nearly to Huntley before the ice cream cones would let us carry on a conversation.

May Day saw me giving *Resolution*'s bottom another layer of 9mm plywood, staggering the serrations so they didn't come on top of the ones in the bottom layer. My visa would run out with the month of June, not quite 60 days away. With so much left to do. The hull needed a fiberglass skin. The bottom needed hardwood keel deadwood. The stem needed its hardwood cap. The hull needed to be filled and sanded fair. The day we could turn the hull over was so far into the future I could not even imagine when it would come.

Dreams pass into the reality of action. From the actions stems the dream again; and this interdependence produces the highest form of living.
　--Anais Nin

Chapter Six

So You're the Guy Building that Boat

Ignorance is bliss. How many times have you heard that? With boatbuilding, it's true. In my mind, I built my Sundowner, *Resolution*, in a year, set out on shakedown trips during the wintertime, sailing north to Fiji, and headed off around the world in October 2006.

Reality set in. Six months after I started building my jig and worktable on September 5th, 2005, I didn't even have the hull completely planked. That's putting in seven or eight hours a day, seven days a week (Sunday mornings off to go to church). So let's say that I work an average of 50 hours a week. At the 26-week stage, I had nearly 1,500 hours put into the project and the hull wasn't even planked. Reality is a bitch.

In New Zealand's May, the leaves start turning red and gold, a subtle reminder that you're living near the bottom of the globe and everything you know is upside-down. Skiing the slopes of Mt. Lyford in mid-July and hitting the beach on New Year's Day.

May 1st saw the second layer of *Resolution*'s bottom applied. Now all I had to do was fill the holes and gaps and hollows with microlite-epoxy mix, sand it smooth and fair, and apply a layer of 10-oz. biaxial fiberglass cloth.

When I put the final pieces of the second layer of bottom planks on *Resolution*'s hull, that meant it was ready for the fiberglass, right?

Wrong.

First I had to round off all four chines and sand them. Then fill all the screw holes with epoxy-microlite fiber mixture. Not an easy job. Each plank got screwed to the chines and stringers with 12 to 15 screws. Let's use 14 as the average because there are more topside planks than chine panel ones. That means I had something like 2,500 screw holes to fill and then sand off flush.

After that's done, the fiberglass goes on, right?

Wrong.

Before that could happen, I had to go over the entire hull with a sharp plane and a sensitive hand, feeling out the high places and planing them down. But even the sharpest plane can't get everything perfectly fair. So after slicing off the high points I marked the low spots and filled them with epoxy and microlite. Finally the entire hull got sanded with a long board and 80 grit sandpaper. That pointed up more hollows that got filled with more globs of microlite.

At long last, the time came to put on the Kevlar and fiberglass. The date was May 16th and the first job was to stretch two sheets of Kevlar onto the hull, running 2m from the stem and 1.2m wide, covering the chine below the topsides. The Kevlar would protect the forward area from submerged and dangerous things like containers floating just beneath the surface, wayward timber, crocodiles, baby whales, and so on.

I cut the swatches of Kevlar and marked their sizes on the hull, then spread epoxy onto the hull with a short-nap roller. The Kevlar went on the wet epoxy, and I wet it down thoroughly with brush and roller. Unlike glass cloth, Kevlar doesn't go translucent when it wets out, so I probably put on too much epoxy. Still, just like the test patch we did on a piece of scrap kauri, the Kevlar stuck fast to the hull without a lot of whiskers and high points. That made it easy to trowel a microlite-epoxy mixture over the edges and across the weave. Next day, I went over the hardened mixture lightly with a little DeWalt vibrating sander to fair it out. It took three more coats of microlite to get the surface smooth enough for the fiberglass cloth, but it got there.

By May, summer barbecues at the club were a thing of the past, but oftentimes in the evening, someone would build a wood fire in the spa pool heater and the open pool back of the clubhouse got nice and hot. I say nice and hot, but around the Waikato Outdoor Society, there's such a thing as "Charlie's Temperature." I like the spa pool at about 40 degrees C while others in the club seem to prefer it around 36. The ideal situation, of course, is for them to have their soak, then put another log or two on the fire, so when I get back from my evening walk, the water is good: Charlie's temperature good.

The spa pool sparked meaningless conversations. Bruce climbed into the spa and the water level went up and over the edge.

"You can tell who's losing weight around here," John said.

"Spending too much time in front of the computer," Bruce replied with his customary wide grin.

"I remember when my little company bought its first word processor," I said. "It was 1982 and the company consisted of three copywriters and an office worker. We got a bank loan and bought a desktop computer, some WordStar software, and a daisy-wheel printer for 2.5 million yen, about $10,000 in those days. The computer had a big 64-byte memory and two slots for 8-inch floppy disks."

"My first computer was a Commodore 64," Bruce said.

John stared out across the lawn.

The laptop I bought to take with me on *Resolution* cost the yen equivalent of about $1,500. It ran for seven and a half hours on a battery charge, had a hard disk of 30 gigabytes, ran Windows software by the handful, and pulled weather charts off the shortwave radio to let me know the sea conditions wherever I was. That's the difference in computers between 1983 and 2008, a mere 25 years. But the boat I was building harkens back to another century, as John likes to base his designs on the 18th and 19th century workboats of Britain. She was put together with modern epoxy, but the wood and the galvanized iron and the bronze in her makeup could easily be of an earlier vintage. Her stocky form, her spliced rigging, her gaff main, her jutting bowsprit all spoke of years gone by.

But in May 2006, *Resolution* had yet to get her protective skin of loosely woven glass fiber called roving filled with West System epoxy. I'd sanded and faired the double planking; all I needed was a few more hands, because even when you're building a boat by yourself, certain jobs need more than two hands and 10 thumbs.

The big day came on May 18th. John Welsford and I donned Tyvek coveralls and set out to epoxy lengths of 10-oz.

fiberglass to the hull. I'd epoxied two sheets of Kevlar to the hull, so it was ready. As fast as I could mix pot of West System goo, John squeegeed it onto the layer of glass cloth laid over the hull. We repeated the process for three days, and on the fourth, with the hull covered in fiberglass and cured epoxy, I went back to working on my own.

As usual, completion of a big job only brought more work.

Loops of glass fibers and the occasional air bubble in the cloth all had to be lopped off with a sharp chisel and then gone over with a random orbital sander. When things get smoothed off, on goes a coat of epoxy mixed with microlite to the consistency of cold molasses. And when that one hardens, another gets rollered over the top of it.

Microlite in epoxy makes an easily sandable layer over the hull; fodder for the long board—a plywood plank 9mm thick, 90mm wide and 900mm long with a handle at each end. I contact-glued a swatch of 40-grit industrial sandpaper to the bottom of the long board, giving it a sanding surface 300mm long and 90mm wide, right in the center of the board. Masked to keep my lungs free of epoxy particles, I set to work sanding the entire hull by hand. As I put my weight on the board, it would bend to make the sandpaper match the curves of the hull, knocking down the high spots and leaving the low points clearly visible. Let me tell you, even a guy with 10 thumbs can become expert with a long board. All it takes is practice, and a 2.5-ton 6.5-meter hull offers plenty of practice. As John so often says: "By the time you finish your boat, you'll have all the skills you wished you had when you started the project." Uhhhh... thanks, John.

My stated objective in building *Resolution* was to sail her around the world single-handed. That's the objective for outside consumption. At the age of 36, I jumped into a new world. For the first time in my life, I was making a living as a writer, a copywriter to be sure, but a writer nonetheless. I suppose it's true in any foreign country, but trying to write English advertising for Japanese clients can be extremely frustrating, and then some. Ever had to make a presentation to a bunch of people who aren't fluent in the language the ads are written in?

Over the years, I exploded once or twice, losing my little company some large clients. The business gradually turned to corporate literature, annual reports, company brochures, advertorials, and so on. I wrote a novel and sent it off to a write-alike contest. Didn't win. The novel went into the bottom drawer.

About the turn of the century, I pulled that old novel manuscript out of the drawer, dusted it off, and started looking for markets. After heavy editing, I sold it to Robert Hale Ltd., publishers in the U.K. They bought my next novel, and my next. But none were best sellers.

I wrote a coffee table book about Japan, and translated another. Still no best seller.

Perhaps, I thought, it's because I've never done anything exciting. That's when John's little Swaggie caught my eye, and eventually *Resolution* was born, on paper at least.

I reckoned that building a blue-water boat by myself would be an exciting undertaking. One that thousands of armchair sailors would want to read. So why was I building *Resolution*? To sail around the world alone, of course, but also to get something to write a good book about.

I wrote in my journal religiously. I sketched in the shop book. I did articles for *Small Craft Advisor, Cruising World, Australian Boat Builder, Practical Boat Owner, Kazi,* and *New Zealand Boating,* and I tried to get ready to write the great New Zealand adventure book. Unfortunately, building a boat is not a very exciting process at all. In fact at times, it got downright boring and took me heaps more time . . . and money . . . than I'd planned on when I started. How in the world could I write an exciting story about putting bits and pieces together to result in a seagoing sailboat?

Ah, but building the boat is only half the story, I told myself. Just wait until we sail. Just wait until we're out on the cold clear water all alone. Just wait. The excitement will come, surely it will.

To keep my hand in, I continued writing Black Horse Westerns. In fact, when John and I dropped by WOS during my March visit to New Zealand, my western novels were still on the club library shelves. They looked fairly well read, and it was satisfying to see them there.

Still, I had to deal with the snail-like pace of the building process. Would it never get done? Would I ever toss the dock lines and sail away on my trip around the world? Would I ever be a hero? Who knows? First I had to get the boat built.

How exciting is it to slather an upside-down hull with a mixture of epoxy and microlite? You get a big scraper and a tub of goop, start at one end of the boat, and work toward the other, scraping the filler down the hull to fill the grain of the fiberglass and smooth out the finish. The next day, you get the random orbital sander out, slap a fresh 80-grit disk on it, fit a mask over your face, and start sanding. Within an hour, you've smoothed off a quarter of the hull and filled the whole workshop with fine dust. In a couple of days, you've got the hull to where you can fill the low spots with goop and take the long board after it to fair the whole thing up.

Remember the long board? That 100x600mm strip of 9mm ply with wooden handles at each end? The one with a swatch of 80-grit industrial grade sandpaper contact-cemented to its face? There's enough give in the ply to allow the sandpaper to follow the curve of the hull, but not enough for it to follow the bumps and hollows. A week of sanding gives you a nice fair hull, but, as John says, *Resolution* is a 10-meter boat—looks great from 10 meters away. Little imperfections are not visible from that far off.

I thought a coat of primer would go on when the hull was sanded off, and John said he'd do some sniffing around for a good wholesale price. Things don't always go as you think they will. The hull stayed in its bald unpainted state, and waiting for the primer; I started work on the keel.

Take a Break, Go to Auckland

As part of John's guest lecturer gig at Massey University's College of Design, Fine Arts and Music, he taught a six-week seminar that required students to design a "round the world cruising yacht for a single-hander."

John invited me to go along with him to Auckland for his sixth lecture. We left Hamilton at 6:30 a.m. and arrived at the university a few minutes before nine. The door was open and the lights were on, but almost no one was there.

"Oh, John," the secretary said. "Didn't anyone tell you? This week is a study week. There are no classes."

John's eyebrows climbed almost to his hairline. "Um, no. Hadn't heard that," he said. "Oh, well. We'll have a cup of tea and go about our business then." He led the way to the lounge area where smells of coffee and strong tea beckoned. The idea was to have a cup of refreshment and go to the harbor to check on a trimaran project John was watching over. After the cuppa, we walked toward the exit, but got only as far as the work lab. "Mr. Welsford. Could I ask you a question?" A young Chinese student, majoring in automotive design but enrolled in the seminar as part of his overall design education, pulled a huge sheaf of papers from his portfolio.

John looked more than happy to oblige. As a matter of fact, we spent the next two hours with seminar students, with John answering their questions one on one.

The design parameters said the boat had to be built of wood, and one student wondered how much weight that would take. John picked up the student's model. "Let me tell you how to do it," he said. "Figure all the weights you know. The weight of the engine. The weight of the ground tackle. The galley. The stores. The fuel. The water. The ballast. Add all the known weights up. Subtract them from your target displacement. That tells you how much weight can be put into the structure of the boat. If you do that and come up with a remainder of something like 300 kg, you know you've either got to make the boat bigger or reduce some of the things you want to put into it." The student nodded like he understood.

From the university, we went to Kevin Johnson's boatbuilding shop. Kevin was constructing a power trimaran to one of John's concept designs. The boat is 45 feet long with extending *amas* (outriggers), and it averages 18 knots under power for 1,500 miles. While another naval architect did the working drawings for the craft, the owner still asked John to travel to Auckland regularly to keep an eye on his boat's progress. Actually, after completion, the trimaran made at least three trips from Auckland to Fiji and back before Kevin sold it.

While in Auckland, we visited The Engine Room, where I bought my 7.5 hpBukh engine. There, they had a carbon-fiber stern tube ready for us, along with the items we'd soon need when installing the engine. Back across the harbor bridge—better known as the Nippon Clip-on—we dropped Quality Equipment Ltd. and picked up a roll of 8mm 3-strand yacht braid that I'll turn into halyards for *Resolution*.

All in all, it was a refreshing day off.

Keeping up the motivation to build a boat that's nearing the end of your original time schedule and not yet half done is difficult, to say the least. Every little task seems to take twice as long as you think it will. You wake up in the morning, take a shower, go for a walk, and come back not yet ready to go to work. So you drive to Cambridge for a cup of lemongrass tea at Fran's and spend 30 minutes reading your current paperback. By the time you get to the shop, check your mail on the internet, answer all the incoming mail that needs it, change clothes, and finally get into the shop, it's 10 o'clock.

You work on the boat until 6 or 7 o'clock in the evening. If you take time off to run to Burger King for lunch, the workday is seven to eight hours long. Even then, after washing up, changing clothes, driving back to the club, fixing dinner of rice and kippered herring, and cleaning up still leaves two hours or more to read that paperback before going to bed by 11 p.m.

That's a day in the life of a full-time amateur boatbuilder—day after day after day. Friday nights were different, because the shopping center in downtown Hamilton stayed open until 8:30 then. So I could zoom down in time to have a meal of good Yorkshire roast lamb, one of superb Indian curry, or one from the Chinese buffet. Late openings on Friday night let me visit Whitcoul's bookshop and browse through the six for $20 paperback table. I usually found six I hadn't read, which set me up for the week as far as reading material was concerned. Filled with delicious food that I didn't cook myself and swinging a plastic bag full of paperbacks, I could meander over to Foodtown supermarket for the few items I needed to live on during the week. The market closed at 9, and if you go just before closing time, many fresh and baked items are marked down significantly. A $10 roast chicken, for example, may be only $6, and a bag of rolls may go for 99 cents. Bachelor living makes a guy more aware of ways to save money. All the money I didn't spend on maintaining life went into building *Resolution*.

All went well, relatively, until July 18th, 2006. That morning, I was trimming kwila on the circular table saw when I cut off the index finger and part of the thumb on my left hand.

Difficulties increase the nearer we get to the goal.
--Goethe

Chapter Seven

Lose a Finger, Gain a Ship

 The day I was using the bench saw to rip a piece of kwila hardwood for the keel was July 18th, 2006. After pushing the work through the saw, I didn't pay attention to how I dragged my left hand back toward my torso. Still running, the saw promptly chewed off my left index finger and a piece of my thumb. It felt like someone hit my left hand with a hammer, but when I looked, my left index finger and a good chunk of my left thumb were missing. Well, not missing, but separated from my hand by the whirling blade of John's bench saw. Surprisingly little blood, perhaps because my blood vessels had clenched with shock. I covered my blood-seeping digits with a scrap of cloth and stepped out of the workshop and into John's design office.

"John?"

He kept his attention on his drawing and didn't turn around. "Yeah?"

"I cut my fingers off."

That got his attention.

 I didn't know it at the time, but John spent his time in the New Zealand Army as a medic. He covered the wound properly, got me set in the front seat of the Vista with my hand held higher than my heart, and went to find my severed forefinger, which he did.

 After rescuing the finger, he wrapped it in a cloth and transported it and me to the emergency center at Waikato Hospital, about 15 minutes away.

 Shortly after we started for the hospital, the shock hit. Cold sweat ran down my cheeks. My breathing turned shallow and quick, almost a pant.

"Be there in a couple of minutes," John said. "Not to worry."

I survived the trip to the hospital, but John rushed in first with my severed index finger to get it put on ice. I sat bleeding in the car the whole time. Nevertheless, despite John's heroics with the finger, the doctors decided it could not be reattached. Being nearly 65 at the time, I suppose the operation would have been more trouble that it was worth, but had I been a lad of 10, I would almost bet the farm that the surgeons would have tried to put the finger back on. Not only could they not reattach my finger, but the surgeon who was to clean up my stumps and suture them couldn't do them that day, so Nurse Agnes gave me several painful shots of painkiller and wrapped my fingers and hand up to wait for surgery on the morrow.

The operation on the morning of the 19th took slightly over an hour, and I woke up in the hospital bed with a bandaged left hand, obviously minus a finger and with a shorter thumb.

I do get on quite well without the finger, though the middle finger sometimes balks at having to do double duty on the word processor and I often drop coins through the hole in my fist. But there was one very good thing that happened because I lost those digits. You see, the money I had budgeted for the year that it was going to take me to build Resolution was nearing bottom. But the insurance payout that came because I was suddenly a Class 4 amputee paid for the rest of the *Resolution* build. As I said above, lose a finger, gain a ship.

I picked possibly the best time in New Zealand history to cut off my finger and thumb. Its unique ACC—an insurance scheme that covers the hospital costs of all injuries caused by accidents and, for working people, covers loss of income while incapacitated or recuperating—took care of all the costs associated with my emergency care, the operation to make presentable stumps of my finger and thumb, hospital ward time, rehabilitation calls, even house calls by the local public health nurses (the stitches were removed by the public health nurse in the clubhouse at WOS as people in club "uniform" walked in and out.)

While I was in New Zealand, my hypertension meds were covered by the national health insurance scheme, but as I left the country, laws changed, and visitors got charged full retail cost of prescription drugs. ACC still covers all people in New Zealand, residents or not, but the laws may change. So if you're contemplating an accident in which you lose a finger or two, you'd be better off to do it soon, while ACC still covers you.

Of course the scars were tender for some time, and I didn't get 100 percent back into boat-building mode for about two weeks. After that, I just had to carry on, using my right hand more than before, and struggling to hold screws and nails in place while I drove them home with pistol drill or hammer.

Under the Southern Cross

As I sat in an 8x10 foot trailer house, my home beneath the Southern Cross. Music issued from the tiny speakers of my laptop, putting my mood to words.

> *Sailing a reach*
> *Before a followin' sea.*
> *She was makin' for the trades*
> *On the outside,*
> *And the downhill run*
> *To Papeete.*

Crosby, Stills & Nash sing.

> *When you see the Southern Cross*
> *For the first time*
> *You understand now*
> *Why you came this way*

Then I hear the Little River Band:

> *If there's one thing in my life that's missing,*
> *it's the time that I spend alone,*
> *sailing on the cool and bright clear water.*

I longed to hear the chuckle of *Resolution*'s wake as she ghosted with yellow topsail spread, lightweight genoa set, and water sail billowing below the vanged-out boom. Why do you want to sail around the world, people ask, and there are many reasons. It feels good to make a long passage on your own, and a circumnavigation is the ultimate long passage. Setting a long-term goal, working step-by-step to accomplish it, and savoring the taste of fulfillment are yet other reasons. Out there on the sea alone, a sailor spends a lot of time discovering who he is, I reckon, and the man who returned to Tauranga would certainly be a different person from the one who sailed away from New Zealand's eastern shore. But in the ordinary scheme of things, the reasons are myriad, and none of them make sense.

Still, one reason stands above them all. With my solo circumnavigation, I'll be adding a major dose of reality to my life.

Day After Day After Day . . .

Here's a typical day of work on *Resolution*.

In the morning, I shaped pieces of kwila hardwood to form the wedge that fits between the bottom just forward of the transom and the propeller shaft log. Then I glued them with epoxy and microfiber putty. Tomorrow morning it'll be ready for another cheek piece to be fitted and glued, and then I'll be able to bandsaw the curve and grind it into a hydrodynamic shape with what John calls a power spoke shave—the edge grinder.

I took the random orbital sander to the hull, smoothing over filled spots that were cured and ready to knock down. Then I got the grinder, the orbital sander, and the vibrating sander out to work on the stem. It needed to be given its final shape, and that takes some care. I lop off the corners with the grinder, shave the sides down with my little Japanese pull plane, smooth things down with the orbital sander, and then give the edges just a little round—so the paint will stick—with the vibrating sander. By dark, I'd finished about half of the stem work.

Soon we'll move the hull out to where we can mount the keel—some 300kg of kwila hardwood. Once that's done, we can roll *Resolution* over. And that'll be a great day.

That said, *Resolution*'s keel was nothing to take lightly. As John drew it on the plans, it's 300kg, with planks 140mm wide and about 70mm thick, that run the length of the boat where possible. When we went to Moxon Lumber to pick the baulks, only a couple were longer than 6 meters. The baulk that would lay against the hull was to be full length. Then halfway through the structure, another full-length baulk would tie things together. Just thinking about building the keel made me apprehensive, so I started with a V-shaped filler section at the stern. It supports the stern tube for the propeller shaft. To get the shaft into the engine bay, I had to cut a big hole in the bottom of the boat. I don't like holes in the bottoms of my boats, but this time the hole would soon accept a carbon-fiber tube and a great glob of thickened epoxy to make it even stronger than the bottom itself. With the wedge in place on the hull, all I had to do was make a log to hold the shaft tube.

Building a log for the tube meant hogging out two 50x140mm hunks of tough kwila. To hog out a plank, you set the bench saw at half the outside diameter of the tube and run it exactly down the middle of both planks. Move the fence so the saw blade is one saw width inside the first cut and lower the blade by 2mm. Run the plank through, slicing a new gouge just to the side of the first one. Turn the plank around. Run it through again, cutting a third gouge to the other side of the first. Repeat the sequence on the second hunk. You adjust the fence and the saw blade until you have hogged a U-shaped groove in each hunk of kwila that's roughly half the diameter of the tube. You should be able to put the tube in one groove and place the second plank over it with the groove down and have the sides of the hunks come together over the tube. Also, it's better if the sawn-out grooves are a little larger than the tube, rather than a little smaller.

Now I had my tube log, and the kwila wedge I'd prefabricated was screwed and glued to the hull in what I hoped was a perfect line-up with the centerline. The top of the wedge (or bottom, if the hull were right side up) formed the platform for the log. I glued one half of the log to the wedge, holding it down with C-clamps and cleats screwed into the sides of the wedge. Lots of thickened epoxy went under the log where it lay on the hull. This was no place for a leak.

While the epoxy cured on the lower half of the tube log, I tried to plane baulks for the rest of the deadwood. Tried. The knives in the Chinese-made thickness planer didn't want to cut kwila. I had to push the hunks through with whatever brute strength I could muster. Believe me, every muscle in the old body aches after a day of wrestling more than 200kg of kwila baulks through a dull planer.

Next morning, I filled the hog in the bottom log with thickened epoxy, smeared epoxy on the mating surfaces, laid the carbon-fiber tube in the hog and pressed it into the epoxy. All I had to do then was prepare the upper half of the log, smear epoxy over the hull where the log would land, and clamp the whole thing together. You wouldn't think it, but the process took me all morning. Building a boat is a slow process, and I was beginning to think that was especially so when building a Sundowner.

Baulk by kwila baulk, I built the keel deadwood. Sometimes I had to use two hunks to get the right length. Sometimes the baulk reached the entire length of the deadwood. It was all tied together with thickened epoxy and clamped down with C-clamps using cleats screwed to the deadwood already in place. As the kwila baulks averaged about 70mm thick, I couldn't use wood screws to hold them in place, so until the temporary galvanized bolts went in, the deadwood was held together by epoxy.

Just a note about *Resolution*'s deadwood.

On March 31st, 2018, John Welsford and I flew to Great Barrier Island, the site of *Resolution*'s demise. Mike Newman the transport driver deposited us at the Mabey Farm upon whose property *Resolution*'s remains still rest. In fact, Scott Mabey offered me the life ring from *Resolution* as well as one of her portholes. I accepted the porthole, with many thanks.

John and I started from the farm house, walked along the loveliest stretch of beach in the entire world, then began climbing up and down, up and down, up and down the endless steep ridges that lay between the lovely beach and the stony fangs in the cove that chewed up *Resolution*.

I went as far as I could, arrhythmia and all, but like so many things I try to do in life, I was unable to make the entire trip. John went on and on and on until he reached the cove and the weathered remains of my boat. He shot photos that included *Resolution*'s Rhode Island registration number, various pieces hull planking, one of which he brought back as a memento for me to treasure. Amongst the photos was one of *Resolution*'s deadwood. Weathered and bleached by the sun, but still together. Ten years of wind and rain and smashing waves had failed to break it apart.

Why Choose a Gaffer?

John Welsford and I discussed my circumnavigation for some months before he sketched Sundowner. Of course, the discussion started with Swaggie and I even bought a set of Swaggie plans. As the idea of going alone around the world gained impetus, I thought of stretching Swaggie by 12 percent to 20 feet. At first, John okayed the idea, but as the discussions continued, he suggested that he design a boat specifically for my solo sail.

John is no novice at designing small boats that cross oceans. As I mentioned, Chris Sayer asked him for a sloop to race in the Mini-Transat. The boat John designed to swoop across the Atlantic with Chris at the helm measured 6.5m LOA with a beam of nearly half that. Further, this swift boat featured twin rudders, a swing keel, and two deep daggerboards forward of the mast. It had so many appendages, in fact, that Auckland's waterfront wags dubbed her "Porcupine." Weird looks notwithstanding, the boat charged to a third-place finish in her first Mini-Transat race, and she still competes.

With all that experience, John could have designed a smashing boat for me to set a speed record for circumnavigations in a 21-foot craft. But his concept sketch was a portly gaff-rigged cutter. A boat that could carry the stores needed for the 6,000-mile leg from Tauranga in New Zealand to Stanley in the Falkland Islands. A boat heavy enough to damp unsettling motion at sea. A boat wide and deep enough to carry a cloud of sail in a breath of wind and dress down to Day-Glo specks of trysail and storm jib in a blow. A boat carefully thought out and designed to carry one man around the world—never mind racing.

Have you ever seen how much sail a gaffer can carry? The working sails are already bigger than those of a comparable Marconi sloop. Remember this: *Resolution* is 6.5 meters from stem to transom. But she's got a battering ram of a bowsprit that adds another meter to her overall length as far as her rig is concerned. A working jib on roller furling rides the end of the bowsprit. A Day-Glo orange staysail hanks to the stemhead stay. And the big main has three lines of reefing points. When the wind whispers, a topsail goes up on its own jackyard. And even more sail can be added by attaching a spare jib beneath the main boom as a water sail. No reason *Resolution* won't move in the slightest breath of air.

Some would say John threw me back into the 19th century with his gaff-rigged cutter design. Not so. Many modern sailors feel that Marconi rigs on fin-keeled boats far outperform gaffers. But gaffers offer advantages of their own.

Resolution is a heavy boat with a full-length keel. She'll be much more comfortable at sea than a lighter fin-keeled craft of similar dimensions. And conventional wisdom says that boats shaped like *Resolution* do better when the weather gets snotty. She should heave-to readily, and she'll point higher while doing it. That gave me peace of mind and I hadn't even sailed her yet.

As I've said before, I could either repair or do without every piece of equipment aboard *Resolution*. The very simplicity I requested helped John decide she should be a gaffer. I looked forward to learning her ropes. Those in the know say a gaff rig is less prone to stalling when oversheeted, and when the wind moves beyond 60 degrees from the bow, that big gaff main comes into its own. Power to burn.

In a gaff rig, the centers of effort are much lower than in a Marconi of the same sail area. That puts less strain on the hull for a given amount of drive. In other words, the gaff rig will be kinder to my boat.

Yes, there are more bits of cordage to a gaffer. No, I won't be able to handle all of the sails from the cockpit. But the wide decks will give me ample foothold when going forward, and each sail is smaller and more manageable than those in a Marconi rig of the same sail area.

My *Resolution* is heavy, displacing two-and-a-half metric tons. And heavy displacement goes hand in glove with a gaff rig.

Turnover is Fair Play

By early August I had completed the deadwood (except for some final shaping that didn't get done until *Resolution* was sitting outside waiting for the crane and hauler trucks to arrive) and the foam plug for the 750kg of lead ballast that would get bolted into a notch in the deadwood sometime in the future.

Time to turn the hull over, do the insides, and put the decks on. Sounds simple, but the whole operation first requires a cradle with which to do the task.

Armed with the usual coin of trade—24 bottles of Waikato Draught—we drove from John's workshop in Hamilton to Bart's Laminated Timber in Cambridge, 16km distant. Sean, the foreman at Bart's, allows us the pick of his reject pile of timbers that failed one test or another. This time we were after material to build a cradle on *Resolution* so we could turn her over. We left with a trailer load of laminated wood. The Waikato Draught went into Sean's worker welfare kitty.

John sketched out the cradle and I built it, sided with massive 50x300mm laminated timbers and connected with 30x150mm planks and fastened with 6mm galvanized bolts. As we planned to roll *Resolution* over on her port side, I rounded all four cradle corners so she would slip right in.

On Sunday August 13th, I had the cradle ready, but one important procedure remained to complete.

Resolution's full-length keel comprises 300kg of tropical kwila hardwood that is so dense it won't float, and 720 kg of lead ballast, which at the moment is a foam plug and several buckets of scrap wheel weights. The kwila was laminated to the bottom of the boat in 70mm-thick layers, but we'd not yet put any bolts through it. Not only that, we didn't have drill bits long enough to make the bore.

I bought two 10mm twist drill bits for the job and John took them over to Ross Todd Motors in Cambridge so Paul could weld a 1m rod to each of them and create the long bits we needed to drive through that tough kwila. We picked up the finished bits just after noon on Monday, zoomed back to the shop, and drilled three holes in the keel for temporary steel bolts to bolster the kwila just in case. Port Townsend Foundry in Washington State was in the process of making *Resolution*'s keel bolts—32 3/8" silicon-bronze bolts of varying lengths. Why so many keelbolts? Because *Resolution* has two sets of them. One set of 22 goes through the floors and the keel deadwood, holding the keel firmly to the bottom of the boat. The second set of 10 goes through the lead ballast to pocket nuts set into the deadwood about 200mm above the top of the ballast. In other words, the bolts going through the lead don't go through the bottom of the boat. During a grounding, then, the keel bolts won't be sprung and the bottom won't leak. Again, in the end, they made no difference.

Keel cinched down with temporary steel bolts and turnover cradle complete, *Resolution* lay ready to take her proper position. John's shop has steel girders supporting the rafters. We shackled two sets of block and tackle from the cradle to those girders—one to turn the boat over and one to act as brake so she wouldn't go over too fast. On the concrete floor, two 2.5m long, 25mm diameter galvanized steel pipes lay beside the cradle. About 2:30 p.m., all systems were go.

John and I put our muscle and weight to the tackle that pulled *Resolution* over. Still attached to her building jig, she protested, then started to tilt. Andrew, recruited from WOS to help, applied the brake on the far side of the boat. Suddenly she reached the point of equilibrium and seemed to balance there. We pulled the tackle a little tighter. Andrew worked the brake. Then *Resolution* slowly went over on her side, where she lay, seeming to pant from the exertion.

Armed with socket and crescent wrenches, our three-man team attacked the bolts holding *Resolution*'s frame crossbars to the building jig. Off with the nuts, knock the bolts through with hammer and punch, and leave a single bolt in the center until last. Then with Andrew on one end of the jig and me on the other, John tapped the last bolt out and we removed the jig—the first thing I'd built when I started the project nearly a year before.

We put our backs to the boat and rolled her across the shop on the galvanized pipes, and then reattached the come-along tackle and the brake. But now *Resolution* was not as heavy as before and the 300 kg of kwila in her keel wanted to swing down beneath the hull. We pulled and she came right along, moving from lying on her side to a full upright position. At last, she looked like a boat.

Soon *Resolution* had the crossbars removed from her frames and the building jig situated at her side served as a platform for materials and a ladder up and over the sheer. I'd be sailing *Resolution* alone, but as I sat inside on a plank set across the quarter berth frames I saw how a Sundowner could cruise four people in comfort with a double berth forward, the head moved to take the place of the navigation table, and the rest of the layout as is. I understand that John already had a double-berth layout in mind, too.

Upright in her cradle, *Resolution* looked impatient. "Get on with it," she seemed to say. "We've got oceans to cross."

Give me a place to stand, and with a lever I will move the world.
—Archimedes

Chapter Eight

Long Time Passing

I revised my internal scheduling. I'd leave New Zealand in October 2007 so as to hit Cape Horn at the turn of the year. It took me a year to get *Resolution* to the turnover point; so the job was half done, right? Wouldn't she be launched and shaken down and ready to voyage in a year? I thought so. I decided so. Then settled down to build the interior of my boat.

Resolution proves that a boat doesn't have to be long to be big. The amazing amount of room inside her hull came home to me the moment I climbed in for the first time. All I had to do was fill that space with cabinetry. I moved the former jig over alongside the boat and built a set of stairs up the side of the hull so I could climb in and out. Still, while huffing up and down stairs every time I wanted to use a saw or a vise might be good for my waistline, it would also waste a lot of time. I reached into the bank account, pulled out some cash, and bought a small bandsaw, which I installed in the galley area of the hull. Now I could do all the small cutting inside the boat, which saved clambering over the sheer every time I wanted to cut a piece of kauri.

I'd completed some of the interior when I made the frames, because they included such elements as the webbing for lockers, the forward waterproof bulkhead, the half bulkhead aft of the chain locker, the two bulkheads that formed the webbing for the mast compression post, the cabin shelf webs, the aft waterproof bulkhead, and so on. Each of the waterproof bulkheads had to be filleted with thickened epoxy and bonded to the hull with fiberglass tape – a time-consuming job in itself. Still, every day I added to the interior, building lockers, shelves, hanging lockers, wastebasket lockers, chart drawers, chart table, folding chart-table chair, galley, quarter berths—I suppose I went through half a dozen blades on the little bandsaw, but it was worth its weight in wood—and time.

The time came to use the beams I'd made so many months ago. The beam at the No.1 frame was already installed, so I could go ahead with foredeck beams at No.2, No.3, No.4, and No.5. Then, moving aft to the No.7 frame, where the cabin aft bulkhead sits, I installed beams on No.7, No.8, No.9 and No.10. That completed the athwartships framing for the bridge deck and cockpit, but I still had laminated beams standing against the wall, one for the cabin roof and two for the quarterdeck aft.

As I completed cabinetry work in the forward cabin, I also laid the stringers for the foredeck. The two outboard stringers, one port and one starboard, were 25x45mm and went from the forward edge of the cabin opening to the gunwale just forward of the No.2 frame. Then came what John called the Princess planks, 30x50mm stringers that marked the outboard edges of the forward hatch, and ran from the forward edge of the cabin opening to the beam of frame No.1. And finally, the King plank. This item, which ran down the center of the foredeck, was actually two pieces of kwila. I used a piece 25mm thick and 150mm wide. On John's big band saw (the little one in the cabin would puke if I ever tried to feed it kwila). I ripped the plank in two to get a pair of planks slightly more than 10mm thick. A couple of passes through the thickness planer got the planks down to 10mm each, ready to epoxy together. Blocking each end up by 20mm, I clamped the center of the planks to the workbench to create a King plank with a bit of a sway in it, to match the sway in the foredeck.

Next morning, the King plank was ready to install, but first I had to put 20mm notches in the beams, No.4 forward to No.1. Beams notched, plank dry fit, drilled, and countersunk, the King plank stood ready for epoxy. I penciled the borders of each beam on the plank, loosened the C-clamps, and slathered thickened epoxy between the lines and into the notches. Starting at Frame No.1, I used my new Black & Decker cordless drill to drive 10-gauge 40mm stainless screws through the King plank and into the beam supporting it.

Time to make the beam that fits the aft end of the foredeck where the stringers, Princess planks, and King plank end.

First, I put a 25mm cleat on each side, up against the No.5 frame and even with the top of the sheer. The first layer of the bonded-in-place beam was a 10x45mm batten of Fijian kauri, glued and fastened to the undersides of the sheer cleats, the stringers, the Princess planks, and the King plank. As soon as the epoxy went off, I laminated two pieces of kauri between each item, starting with sheer to stringers. In the end, the fillers stood proud of the stringers and planks, but a sharp plane made short work of bringing the laminated beam to the same height sheer to sheer.

John drew a deckhouse with sides parallel to the sheer. I decided to take the easy way out and make straight doghouse sides. I put the aft end of the carlings 300mm from the sheer and the forward end 400mm in, creating a cabin wider aft than forward. Then I half dovetailed supports every 450mm along the carlings to take the side decking.

Now for the tough job.

ADOS puts out an epoxy primer that incorporates a strong fungicide. John said we would put three coats of the stuff on the inside of the boat. Yuk.

I bought a couple of cheap garden weed sprayers. Then, dressed in Tyvek suit, booties, surgeon's gloves, and a special mask designed to keep toxic fumes from entering my pulmonary tract, I mixed a liter of resin with a liter of hardener and poured the stinky stuff into the sprayer. I started at the bow and sprayed every inch of the wood inside the boat. By the time I'd reached the stern, I'd refilled the sprayer twice with a bit left over. In between coats, we tested the mixture on a stray piece of kauri. It penetrated a full 3mm, the most I have ever seen epoxy sink into wood.

Next day the second coat of primer went on the interior, and another coat the day after that. The kauri visible inside turned a beautiful golden honey color. With 100 percent hindsight, if I were to build *Resolution* again, I would use wood flour as thickening for fillets and fillers. That way the seams would be nearly the same color as the primed kauri wood in the interior, which would minimize the need to paint.

About the time I got the interior primed, I discovered the Rialto cinema complex in Hamilton. After that, I treated myself to a movie a week (if there was a movie I wanted to see). I probably saw more movies in New Zealand than in the entire 30 years I lived in Japan.

Paul Newman had recently died and I'll never forget the last movie I saw that featured him—the animated film called "Cars." Newman was the voice of the Judge, a Hudson Hornet that had once been the toast of the stockcar racing circuit. An entertaining movie with a stellar vocal performance by Newman. Lots of "in" jokes that only people well acquainted with the world of cars would understand. Me? I started working on Toyota advertising and marketing materials in 1979. Since then, I've written corporate ads, brochures for every line of cars Toyota makes, including Lexus, done detailed brochures on new engines, aerodynamics (Toyota was the first Japanese automaker to invest in a full-sized wind tunnel), suspension systems, and a whole bunch more. Then, on the side, I did a whole series of advertorials for Newsweek magazine that focused on the Mitsubishi Pajero and the annual Paris-Dakar Rally. Further, I did vehicle reviews for 4-Wheel Fun, a German off-road magazine. Somehow in the process, I managed to learn a bit about cars. So I got most of the jokes in the movie, but a few still went over my head. John Welsford, a former racer and automobile enthusiast, got *all* the jokes.

Once the interior was primed, every piece added thereafter also had to be primed. As I readied to lay the fore and side decks, for example, I had to prime one side of the 6mm plywood with three coats of that awful-smelling and highly toxic primer from ADOS.

Resolution was an equal opportunity boat. Not only did she have both King planks and Princess planks, but forward she had a Samson post and aft she had two Delilah posts, which are somewhat smaller than Samson posts—naming courtesy of John Welsford.

The Samson post measured 70mm square where it poked through the King plank and stood about 120mm above the deck. Just beneath the King plank, it became a hunk of kwila 70x140mm. I dry fit it and drilled a 10mm hole through the post, the bulkhead, and the 25mm packer behind the bulkhead. Then, I slathered epoxy on every surface that would contact another surface—the hole through the King plank, the underside of the King plank, and the aft face of the bulkhead. Epoxy went on the post as well, and then I tapped it up through the hole, bolted it in place, and then drove two 10-gauge x 45mm stainless screws through the King plank into the notch in the Samson post. More screws went through the No.1 bulkhead and into the post, staggered on about 100mm centers.

Then, as soon as the epoxy cured, I went to laying the foredeck and side decks.

For some reason, perhaps experience, the foredeck went faster than the hull. Starting about the 20th of December, I had the first layer of 6mm decking laid by the 26th, and the second layer (laid at a 60 degree slant) less than two weeks later. By mid-January, all the permanent stainless screws were driven through both layers of decking and into the beams, stringers, Princess planks, and King plank, and the screw holes were filled with epoxy and microlite and sanded smooth.

Just a word about screws. In New Zealand, John swears by self-tapping screws from a local aluminum company. True, the price can't be beat. But the screws have little ability to pull work together and even No.2 square drive screws seem prone to distortion. I found common stainless wood screws from the local DIY house, PlaceMakers, which cost a little more than the self-tappers, worked best for me. When I switched to them, I lost almost none to distortion, and they pulled themselves down into the plywood so I didn't have to stop and ream out special countersinks for them. Maybe that's why the foredeck planking went faster.

A boat with a deck is even more like a real yacht than one that's newly turned over. But as I stood back to looked at my handiwork, I realized a year and a half was almost gone and my "boat" didn't even have a cabin or a cockpit or an engine... there was so much left to do. Was I even halfway through?

The days are long and the nights short when the year turns in the Southern Hemisphere. The sun rises about 6 a.m. and sets at close to 9 at night. Stores don't stay open any later. Companies keep the same hours. So basically the whole country has five or six hours of daylight in which to do whatever.

The early months of the year are sailing months, too, and many of New Zealand's open-boat cruisers take vacations that include the Christmas-New Year break and extend for two or three weeks beyond. Auckland is known as the City of Sails, and has more pleasure boats per capita than any other city in the world. The Hauraki Gulf extends eastward from Auckland and provides wonderful cruising grounds for ubiquitous dinghy camper-cruisers.

One place the wooden boat cruisers congregate is Mahurangi near Scot's Landing. There I met Dave Perillo, intrepid sailor and owner of *Jaunty*, one of John Welsford's Navigator dinghies. Navigator is just under 15 feet long and most often rigged as a yawl. When I say intrepid, I mean that Dave shipped his Navigator to Fiji and then spent three months sailing those islands, often as far as 50 miles offshore, and de-stressing from a hectic life ashore. I set myself the goal to sail around the world in a small 6.5-meter cutter, but the boat weighs three and a half tons in full cruising trim. Dave's little *Jaunty* comes nowhere near a ton, even loaded to the gunwales with water and stores. Still, according to Dave, three months spent gunk-holing around Fiji is the ultimate life for the freedom-loving open-boat sailor.

A Bit about Navigator

John says Navigator turned out to be a surprise. Designed as a race trainer for a local Auckland club, its uses and legacies have just kept on expanding. First a customer came looking for a long-range cruising dinghy, and liked Navigator's hull and internal layout, but wanted a different rig. Ergo, John bequeaths Navigator with yawl rig that keeps the boat underway and under control when the wind pipes up and the mainsail must be dowsed. The request for cruising dinghy meant enough storage space for a week's travel, plus water for the same, and John designed each locker to have a watertight lid or hatch so they would act as floatation in case of a capsize, allowing the crew to right the boat and get on with it. John says, "In fact when we tested a later boat, she supported her crew comfortably and was easily bailed out to get back underway."

The client built his Navigator in about four months, in a single car garage. Named *Ddraigg*, the boat went cruising, raced with more success than expected, and of course was daysailed constantly. Says John, "All in all, she did more mileage than almost any open boat I've seen."

Let's listen to Dave Perillo about how he came to own a Navigator.

"I still had a 10m catamaran," Dave writes, "but I just knew I needed something else, and there was a small ad complete with a grainy black and white photo of a Navigator for sale in Wanganui. Three days later that boat was mine and I'm cruising down the harbor here in Auckland having a ball. Three weeks later I was cruising Fiji, shattering the typical offshore cruisers' image as to which is the perfect boat for tropical voyaging."

Dave lists five reasons for owning a Navigator.

1. Comfort—both underway and swinging on the anchor at night.

2. Almost no fuel requirements. Sails and oars if you wish. Dave's Navigator *Jaunty* had a 4HP outboard, but 2HP would be plenty, he said.

3. Trailerability—*Jaunty* lived under cover in Dave's garage. "When I'm ready to go," he writes, "I just hitch her to the car and I'm off." Very cool.

4. Affordability. Considering the high prices of so many fishing and speedboats, Dave says John Welsford's Navigator is positively cheap. Also, "in a Navigator," Dave writes, "your summers will be most memorable, you will be the envy of everyone, and you and your family will learn a bunch of new skills."

5. Adventure. Dave proudly declares: "You can pretty much go anywhere a big boat can and many, many places a big boat cannot including small boats with big engines. My Navigator will sail in just over a foot of water off the wind. I can explore rocky coasts at unbelievably close quarters. She surfs well and carries a fair load. I am a keen fisho and I have caught many great fish from her; whether trolling lures or bottom fishing she's a perfect platform. And at the end of the day when you end up miles from home as happens often I can set the boom tent in a jiffy. From the dry storage area under the foredeck I'll grab a foam mattress, a sleeping bag and a big feather pillow I stole from home and sleep in total comfort. Sleeping on any boat is usually pretty comfy but on the Navigator it takes on a whole new ambience."

Where Dave touts cabinless cruising, my *Resolution* needed both a cabin and a cockpit before she could come close to crossing oceans. That said, the boat must also have an engine, and small though the little Bukh is at 62kg, it had to go in before the roof went on. Not that the engine can't be installed or extracted with the cabin on the boat, it's just a lot easier with only the cockpit framing in place and a sheet of plywood laid down as a floor.

I sat in my "office" aboard *Resolution*, surrounded by the mellow honey color of raw kauri and the scent of fresh-cut wood. My college-ruled notebook lay at a slant on the chart table, which was more than large enough to take a full-sized chart folded in half, as it measured 870x560 mm. And there was a small triangular space further outboard of the table itself. At that moment, it was full of nails and screws, a measuring tape, drill bits, a Japanese saw, and my $4 hammer hanging from a divider.

To the left, a top-loading locker expanded the usable space of the office. In port, materials could be spread out—reference books, printouts from the web, research notes, pens, pencils, erasers, sandwiches, and Pepsi MAX—and there would still be ample room to write in longhand, or to dig out the trusty Apple iBook laptop and bang away at the keyboard.

How many 21-ft boats do you know that come equipped with an office?

A Big Little Boat

With *Resolution* standing upright, the first impression was "My God, that's a big little boat!" And six weeks after the turnover, she was still a big little boat.

As I said, I could sit at the chart table on a fold-down seat that stayed out of the way against the wall of the galley when not in use.

The office is slightly forward of the boat's center of buoyancy, but still positioned in the low-motion area. To the right, a quarter berth stretched all the way to No.9 bulkhead, nearly three meters. In other words, someone could be sleeping in the bunk and another person would still have room to sit at the forward end on either side. Not important during a solo circumnavigation, but vital when there are two or three aboard. That lets the crew brew some coffee, make soup, or plot a position without waking a sleeper.

To the left, forward of the chart table, two large lockers lined the hull, port and starboard. The port locker was lidless, ready for bags of sails or mesh bags of supplies. Also, just aft of the big port locker stood a hanging locker with additional stowage in the bottom. The large starboard locker had two levels, each with an access lid, so supplies could be stored according to frequency of use.

The two lockers formed the boundaries of a 90x90 cm (three feet square) space for the head. A Porta Potti sat on a lidded riser that tripled as platform for the toilet, storage for toilet supplies, and a step up towards exiting the forward hatch.

This head area was located directly beneath the forward hatch, which meant stand-up room in which to pull up my pants. And, as the head was an in-port facility, meaning you only use it while you're anchored or tied up in a harbor, there was plenty of room for a spit bath while sitting on the closed Porta Potti.

Still, the head was not the forwardmost part of the boat. No.2 bulkhead, at the forward end of the head area, was a half-crash bulkhead glass-taped to the hull on both sides, watertight to about a foot and a half above the waterline.

The chain locker was located between No.2 and full crash bulkhead No.1. It slanted up toward the stern and could handle 50m of chain or more. After the deck was laid, I fit a bronze chain pipe and led a 2-inch wet exhaust hose from the chain pipe to the locker. The rubber hose would keep the chain from rattling with the wave action. If there's anything worse than the clanking of halyards against aluminum masts, it's anchor chains rattling in their pipes.

The area above the chain locker was left open for sails and a row of pegs down each side of the hull gave me a place to hang cordage. Also I fit a reel in that area to hold 300 feet of one-inch nylon rode for use with the No.2 Danforth anchor and my 9-foot para-anchor.

There was even more room forward. Watertight crash bulkhead No.1, also glass-taped to the hull, had a 280x280mm access hatch (watertight, of course). The upper part of the vee in *Resolution*'s bow forward of No.1 bulkhead was the anchor well. Beneath, accessed from inside through the hatch, lay storage space for awnings, dodgers, lee cloths, flags, and other lightweight items.

That's not all. Starboard and port lockers under the deck offered even more places to stash gear. Access was through large open ports, but I fit netting to keep the contents from falling out.

Starboard, the 'neath-the-decks lockers extended aft, crossed the office space, and went all the way to the cabin's aft bulkhead. In the main cabin area, above the quarterberth/settee, seat backs also added storage space.

Did I say that the chart table hinged upward to reveal a compartment for charts folded in half? Another locker beneath that held various gizmos and gillhickies, and yet another was set up to hold a trash bag.

Of course, all six of the berths/settees also had storage compartments underneath.

I'd not built the galley section yet, so I won't wax specific about its spacious wonders. And the area aft of the engine was only bulkheads and a few stringers. The No.9 crash bulkhead was filled flush to the hull and ready for glass tape to be epoxied around the edges, fore and aft.

Such is the enthusiasm of the homegrown builder, working one tiny step at a time, hoping to make a dream come true.

Democracy does not guarantee equality of conditions—it only guarantees equality of opportunity.
--Irving Kristol

Chapter Nine

More Power to It

In between laying the foredeck and setting up the cockpit framing, I readied the engine bay for its occupant.

The shaft tube protruded just the right distance from the bottom of the boat to the transmission. From the remainder of the hogged out tube log, I made a short log to fit from the inside bottom of the boat to a point 100mm from the end of the tube. I slathered the upper half of the log with thickened epoxy and clamped it to the top of the tube. My hands covered with surgical gloves, I used a finger to smooth the glops of epoxy that extruded from the sides and end of the log, creating smooth fillets and removing any excess because the lower half had to go in after the upper part solidified. Actually, the next day, the lower half of the hogged log slid neatly into the space I'd left for it, and I soon had a solid log protecting the portion of the shaft tube close to the bottom of the boat. What's more, the hole in the bottom, now generously filled with epoxy and microfibers, was probably stronger than the original.

I built a platform for the 61-liter fuel tank above the shaft log and back up against the No.9 bulkhead. Just forward of the shaft pipe came the big 70mm kwila floor of the No.8 bulkhead, with a similar floor on the No.7 bulkhead 600mm or so forward of that.

I'd paid a king's ransom for a set of bronze keel bolts from Port Townsend Foundry. Well, to be honest, not a king's ransom . . . a minor prince's, perhaps. Normally, the bolt holes would have been sunk while the boat was still upside down, and the bolts cinched home. But the keel bolts had not arrived, so I sank holes for three temporary bolts, two in the engine bay floor and one in the forward cabin. Threaded steel rods held the deadwood in place, and we rolled the boat over.

Procrastinator that I am, even after the bolts arrived, I worked on other parts of the boat. Lockers took shape. Beams appeared over the forward cabin. The King plank went in, along with princess planks and stringers. Cabinetry for the galley. A chart table. A folding seat for the "office." I even built a small locker to hold a trash bag. A platform locker for the Porta Potti. And the floors that formed the backbone for the engine logs. I notched the floors to receive the logs, then ran a string through the tube. A round wooden disk with a hole in the center plugged the aft end of the tube and located the string exactly in the middle. I stretched the string to the galley bulkhead and held it in place with a small nail. That string down the shaft tube's centerline helped me make sure the engine logs were in just the right place, fore and aft as well as athwartships. I screwed them down with 8mm lag bolts.

But there was still work to do on the engine bay. In effect, it had to be a watertight box to enclose both the engine and the fuel tank—lined with fiberglass cloth all around, even the floor, and covered with three coats of West System epoxy, then two coats of primer and three of high-gloss exterior enamel. At last, ready for the engine. Almost. I still had to sink keel-bolt holes and put in the bolts. And I had to finish that job before the engine could go in.

Out came the big Hitachi 1/2" drill. Out came the meter-long drill bit Paul down at Ross Todd Motors in Cambridge made up for us. Starting in the stern, I bored the holes, taking my marks from perpendicular lines I'd drawn on the major bulkheads as I was making them so months ago.

Then I came to the No.2 bulkhead, which formed the aft end of the chain locker. But its hardwood floor lay forward, and the chain locker prevented access with a meter-long drill bit . . . or with a 300mm length of bronze bolt, for that matter. Now what?

There was 300kg of kwila keel deadwood under that No.2 floor. It's a crash bulkhead, with a 9mm plywood web covering the bottom half. Early on, I used 200mm pieces of biaxial fiberglass cloth to tack the bulkhead to the hull on both forward and aft faces. The floor for the bolt holes I couldn't drill attached to the forward face of that bulkhead with 40mm stainless screws placed in a zigzag pattern every 75mm.

Decision: put a second floor made of kwila hardwood on the aft face of the bulkhead. I cut it 70mm thick and 150mm wide, and carefully make sure it fit into the available space. I slathered it with epoxy-microfiber mixture and fixed it in place with three lag bolts through the bulkhead and into the original floor, and one lag bolt through the floor and into the keel 35mm starboard of the centerline. Nothing short of a steamroller would move that floor.

The meter-long drill had no problem boring the keel bolt hole dead center, and all was ready for the keel bolts . . . except for one thing.

Resolution sat too close to the concrete floor. I couldn't get under the keel with my 28mm drill bit to countersink the bolt holes.

How do you raise a half-finished boat? With the same jack you use to change a flat tire. The one from my 1987 Mitsubishi Mirage lifted *Resolution*'s stern far enough to lever up the aft edge of the cradle and block it in place with a couple of 2x12s. Same process with the bow, and *Resolution*'s keel stood a good 60cm off the floor. Then I had no problem with the counterboring.

Ready to set the bolts. Almost.

Remember the ADOS Timber Primer we used on the interior? The fungicidal stuff that penetrates soft wood up to 5mm? Well, before setting the bolts, I pounded a wooden plug in the bottom of each bolt hole and poured a liberal dose of the primer in the top. It soaks into hardwood well, too.

Time to set the bolts.

Nut and washer went on the upper end. The bolt got coated with a blend of West System epoxy and microfiber and twirled slowly as it sank into the hole. The excess first shows in the top counterbore. With the washer and nut solid on the floor, check the keel. Protruding just short of the bottom. Good. Washer and nut go on and the nut is run as far up on the threads as it will go. Now. Back up in the boat, each upper nut is cinched down tight, fixing the keel bolt solidly in place. Once the epoxy sets, nothing short of a coral reef in a hurricane can move that keel.

With the keel bolts in place, we could install the engine.

Charles T. Whipple

Small Vacation in Plimmerton

In late March, I was invited to Plimmerton, near New Zealand's capital of Wellington. Each year, the Plimmerton Boating Club hosts the Plimmerton Classic & Wooden Boat Show. There's a huge ramp where members launch the boats they bring along on trailers. Unlike Japan, New Zealand requires no special driving license to pull a trailer, and every car sports a tow bar, even my little Mirage.

March 23rd, a Saturday, I left my little trailer house just after 6:30 a.m., and hit the long road to Plimmerton, a small town on the coast of the Tasman Sea, some 550 km south on State Highway 1. I followed the main road along the shoreline. The bay glistened a dark blue, a sign of deep water and a rocky bottom. I knew I was in the right place when I saw the distinctive yawl rig of a John Welsford Navigator, beating against the freshening onshore breeze. After a seven-hour drive and two pee breaks, I turned in at the gates of the Plimmerton Boating club slightly before three in the afternoon.

When I arrived from Hamilton, most of the 23 boats entered in the show were lined up on the tarmac. They ranged from a 6 Metre Whaler down to a 4-foot lapstrake dinghy for kids. John Welsford's boats were well represented with three Navigators, a Houdini, the 6 Metre Whaler, and a fat little Truant called *Hookey*, built by Jim Shaw to meticulous yacht standards.

I blundered about the clubhouse until I found event chairman Richard Schmidt's wife Sonya, who steered me toward coffee and cake and a fine view from the second floor balcony. Richard's Navigator *Bootstrap* owned the set of sails I'd seen earlier.

I was the featured speaker at the dinner that night, and told the assembled boaters about progress with *Resolution* and my motives for wanting to sail around the world by myself. Of course they wanted to know how much *Resolution* was going to cost to build, and of course, me being far over any ideas of budget I had when I began, failed to be specific about the amount. I didn't keep meticulous records; like a compulsive gambler, I just couldn't admit how much building *Resolution* was costing me. I'd arrived in New Zealand ready to whip out an ocean-going yacht in six months and be sailing her before the year was up. There was so much I didn't know, and no one disabused me of my unrealistic expectations. I just had to do the work one day at a time and realize each day that it was another day longer than I had anticipated.

Richard and Sonya, intrepid South African immigrants to New Zealand, put me up in their guest room for the night, and we walked the neighborhood with their dog and daughter Tracy, who had the most amazingly wild head of hair I've ever seen on a youngster.

The Plimmerton Boating Club sits on a finger of old lava sticking out into Poirura Harbor forming Karehana Bay. The morning of the second day, Pete launched the 6 Metre Whaler and we took a sail, five adults aboard, though the boat could have easily held two or three more. In the fresh breeze, we headed for Mana Island across the harbor. With four hefty men on the windward seats, the lee rail never came near the water. With its long waterline and slim, flat bottom, the whaler skimmed over the waves, seeming just to slide along the crests without bobbing down into the troughs. An exciting, exhilarating ride in a well-designed and well-built dinghy. John says he designed it for a crew of cadets in Australia, but it went well with a crew of middle-aged men. A better ride would be hard to imagine.

The 6 Metre Whaler

You can get the whole story from John Welsford's jwboatdesigns website, but let's just pick and choose some of the highlights about the craft. Here's what John has to say:

"A friend told me that he was keen to establish a boat-based outdoor youth training course to help change the ways of some of the local kids and was considering building a pair of brand new Naval Whalers. Now the cost of a Naval Whaler is staggering and way beyond any budget he could raise, so we sat down and worked out the key features of the clinker-built 27-foot Naval Trainer and I went home to translate those thoughts into a form both affordable and buildable by volunteer labor.

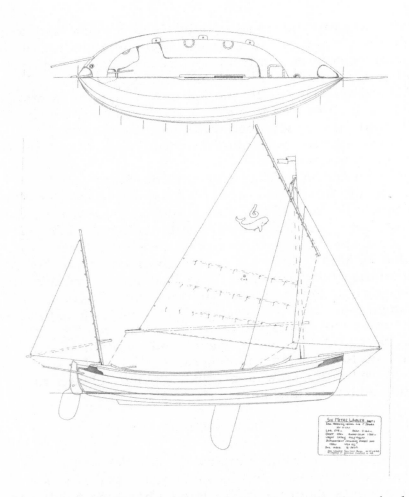

"After a few months of phone calls back and forth and a bit of work at the drawing board, the 6 Metre Whaler emerged. It used the very successful yawl rig popular on the Navigator cruising dinghy, and the same simple plywood lapstrake over web frames and stringers structure as well. Traditional in her appearance, double ended, with two masts, and enough sheer to be really pretty, she needs only about NZ $4000.00 worth of materials and about 220 hours of (competent) labor to complete a very nice craft that really looks the part.

"She has a pair of straight alloy tubes as masts, the same sprit booms and proportions of sails as Navigator. There are three rows of reefs in the main, and if the weather requires, she will still balance under main only, or under jib and mizzen only, which gives lots of choices as the wind increases in strength. All spars fit inside the boat to reduce windage when rowing and simplifying stowage when trailering.

"An outboard motor hides in its well under the aft deck, and the steel centerboard makes the boat well-suited to the shallow harbors and estuaries where she will be based. But that 'board' – 85kg of steel – will steady her under what is in fact a powerful rig for her weight. Responses from owners say she is a very fast boat under sail. And if worse comes to worst, there is enough buoyancy built in to keep all of the occupants afloat while she is bailed out.

"As a daysailer, one or two can handle her easily. I like these big open boats for the sheer feeling of space and comfort available in a relatively small boat, and it is in this role that the 6 Metre Whaler is most popular. One small group is chartering a couple and several are being used very successfully as cruising dinghies."

Charles T. Whipple

Little Red Bukh

The day before we installed the Bukh, John and I lugged the motor outside and put it on the blocky little table John keeps in the shop yard. I sprayed the entire engine with degreaser and washed it off with the garden hose. How it sparkled!

It spent the night outside, wrapped in a ground cloth to keep the rain and dew off.

Sunday morning. A day of rest, but also a day for putting engines in boats. At a few minutes after eleven, with the sun shining in blue skies, we picked up my 62kg Bukh, hauled it into the shop, shuffled around the bandsaw and the joiner, positioned ourselves at the stern, and on the count of three lifted the engine onto protective sheets of fiberboard on the floor of the cockpit. We then slid the Bukh forward to the bridge deck framework, which had a piece of white particleboard laid on top, and lifted it onto the board. John held the motor while I positioned myself in the main cabin, then I held it while he moved. In position on each side of the engine, we simply lifted it up and over, and then gently set it down on its mounting logs.

"OK, it's all yours," John said, and disappeared into the office.

I grabbed the engine and wiggled it on the mount logs. It lined up with the shaft hub athwartships but was 5mm low. I cut two 100mm pieces of 5mm kwila and slipped them under the aft ends of the engine's mounting rails, raising the drive by 5mm but leaving the forward end at the same level. I slid the engine aft and the flexible connector lugs fit perfectly. I cinched the nuts down, put the transmission into neutral, and turned the shaft, propeller and all. Excellent fit. After pulling the engine forward a bit to ensure enough shaft space for the stuffing box, I drilled for lag bolts and fixed the engine in place.

Now all I had to do with connect all the peripherals.

The fuel line attached to the 61-liter fuel tank in the aft compartment of the engine bay. Its loose end lay over the top of the engine. I trimmed it to the proper length, and pulled the plastic cap from the line leading to the fuel pump. The odor of diesel filled the cabin. I put a hose clamp over the fuel line, shoved the line on the nipple, and cinched the clamp down with a small Stanley socket wrench. The remainder of the fuel line hose formed the return from the injector overflow back to an inlet in the fuel tank. One system finished. Water, exhaust, and venting left to go.

I'd installed *Resolution*'s only thru-hull fitting when I was preparing the engine bay. It was located on the port side of the engine, inside the watertight bay. The 1/2" thru-hull fit a Marelon sea cock, from which a 1/2" line ran to a water filter, from there to a T-joint and then on to the engine and the galley. Hoses, fittings, and clamps are all you need.

The exhaust system posed a different problem. At first I thought of putting the waterlock in the cargo hold aft of No.9 bulkhead. But when I placed it there temporarily, I could see that it put too much exhaust pipe too high just aft of the exhaust elbow. I put the waterlock alongside the engine, but off to the starboard where the toolbox was originally to reside. But that was OK. The exhaust system was shipshape and Bristol fashion.

Venting was simple. Tube over the fuel tank's vent fitting. Vent out the forward wall of the cockpit under the bridge deck overhang. Everything was ready, except the cockpit wall.

I'd soon have to deal with the electrics, too, but I had the battery boxes made and would soon fit them in place. The starter battery went next to the engine and the house batteries were just forward of the galley.

Spectators Welcome

PJ drove a truck, a big Mack that grossed 30 tons, tractor, trailer, and load combined. That day he sat on *Resolution*'s foredeck, his legs swinging in the still-empty main cabin. "Jeez," he says. "If a man had never built something himself, he wouldn't be able to see all the man-hours that have gone into this boat."

I couldn't help smiling.

He thumped the foredeck with a big fist. "Solid," he says. "Brick shithouse solid."

It's amazing how two layers of 6mm plywood, saturated with ADOS timber priming epoxy and laid on 60° angles to each other turn out. The epoxy primer not only protected the wood from fungus, but it also stiffened the 200mm-wide planks to a surprising degree. Then, sandwiched between the layers, West System epoxy with microfiber thickener. The result was a deck you could dance a jig on, as John would say.

After PJ and his girlfriend left, I sanded a round on the foredeck sheer edges so 100mm biaxial tape could wrap smoothly over the sheerline and down onto the topsides. It actually took longer to sand the round on the foredeck planking than it did to apply the tape, but then the foredeck was taped from the forward edge of the doghouse to the anchor well in the bow. What's more, a 200mm swatch of 10-oz. cloth ran down the center of the deck, lapping the king plank on both sides. Epoxy and microlite filler'd been applied and sanded, feathering out the edges of tape and center swatch. She was ready for a layer of fiberglass cloth.

Night and Day

At night, I planned. A shakedown cruise to Nelson just to make sure everything worked as designed. Depart for the solo circumnavigation in late October so as to round Cape Horn in the dead of summer, January 1st, if possible. Later I found out that early March, specifically between the 7th and the 14th brings historically placid weather at the Horn. Next time, if there is a next time, I may change my route.

At that point, here's how I decided to go. A long leg from Tauranga, New Zealand, to Stanley in the Falkland Islands. Rest and recuperation. Sail from Stanley to Hout Bay, just outside Cape Town, South Africa. Spend some time with Roy McBride and get ready for the next leg. Depart Hout Bay, round the Cape of Good Hope rather close inshore, then bear off for my next destination, Fremantle, Australia. Rest, recuperation, boat maintenance, whatever. Leave Fremantle, sail the Great Australian Bight, round Tasmania to the south, cross the unruly Tasman, round the North Island of New Zealand and sail down the eastern coast to my departure port at Tauranga. Not an official circumnavigation because I'd not cross any antipodes, but a sail around the globe anyway. I pored over the pilot charts and the route map I bought off the internet, and started collecting the paper charts I would need for the voyage.

Ship of Dreams

    By day, I built the cockpit. Bridge deck and cockpit seats, smelly locker for paints and thinners and other things that give off noxious fumes right aft, big cockpit drain tubes that led through the smelly locker, access lid to the cargo hold through the cockpit floor. Access to the fuel tank also through the floor. Slowly, far more slowly than I wished, *Resolution* took shape. I thought I was getting closer, but still didn't realize just how far away I was.

    The stern had no transom. The middle of the boat had no cabin. The forward anchor well remained unfinished. None of the fiberglass on the decking was laid. Although I had a great deal completed, another great deal remained. You've heard of amateur builders spending decades on their dreamboats? Believe it.

When you have a dream, you've got to grab it and never let go.
    --Carol Burnett

Chapter Ten

Uppers and Downers

It's hard to keep motivated. You hear stories of people who spend 10 years building a boat. But they're holding down a day job, too. Ralf Schlothauer visited. He was building a Penguin of John's design over in Te Aroha, about 30 miles away. He worked full time and spent about four hours a weekend on his Penguin. His work was meticulous, and his boat was to be a definite source of pride. But between his job, his wife, and his two small children, I'm sure he simply couldn't wait to get back out in the garage to work on his Penguin.

I didn't have a job. My family was scattered from Hell to breakfast, literally. But, for some strange reason, I couldn't stay focused.

Rebecca Hayter, then editor of *Boating New Zealand*, the country's premier boating magazine, wrote an excellent book entitled *Oceans Alone*. She followed the single-handed racing career of Chris Sayer, a professional boatbuilder, as he built a Mini-Transat 6.5 to a John Welsford design, sailed it in races and shakedown voyages around New Zealand, took it to Europe, placed third in a Transatlantic single-handed race, sold the boat, returned to New Zealand, built another boat, watched it sink out from under him, built yet another mini-Transat 6.5, and got married. "I've never seen a man so focused," John said of Chris. He'll never say the same of me.

An Apple a Day

Jean waited for me at Newstead Orchards. In days gone by, 10 years intervening, she managed the shop at the orchards, where tree-ripened apples of several strains sat in plastic bags. And at $3.00 for a kilo and a half of fruit, who wouldn't go to the orchards once or twice a week?

She looks the same, though we both have aged a decade.

"Oh, Charlie," she exclaims, holding out both hands to clasp mine. "It seems like just yesterday. What will you have today?"

I liked the feel of her hands clasping mine. It made me feel like a Kiwi. How could I resist? I toured the tables full of tree-ripened fruit.

Royal Gala, Braeburn, Pacific Rose, Pink Lady, Red Delicious—all picked that morning (I know I belabor the point, but it's important). All perfectly ripe, crispy and full of juice. In fact, if you wanted fresh squeezed apple juice, there was plenty, in your choice of one or two-liter bottles.

My eyes fixed on the Pacific Roses. Should I buy the kilo and a half bag for $3.00? Or go for the three kilo one at $5.00?

I took the latter, as I had so many times a decade ago. John grinned. "None better," he said, and picked up his own five kilo bag to take to Denny's place, where we crunched those fresh sweet apples in lieu of desert after her sumptuous meal.

Returning to New Zealand cemented in my mind just how many really good friends I'd made in the 30-plus months I'd squatted there, first at WOS, then in Newstead, near the workshop where *Resolution* slowly, oh so slowly, took shape. The second trip reminded me just how precious those friendships are.

Take a Break

While I was in New Zealand as a full-time boatbuilder, three of my daughters had babies—two boys and a girl. "Time to come home for a week or two," the girl e-mailed me. "Nanna and Emma will come to Japan from the States, Tina and Eve live close by, and you can meet Hibiki, Taisei, and Nanase. How soon can you get here?"

I'd gone to New Zealand with a one-year open-ended airline ticket, so if I decided to fly back, I had to rearrange my own flight schedule. After receiving orders to "take a break," I wasn't able to get a plane out of Auckland for almost two weeks.

I boarded the plane with toy trucks for the boys and a rag doll for the granddaughter, and settled back for a 12-hour flight to Taipei. Four movies and three meals later, the plane made a faultless landing, and I ensconced myself in the transit hotel for the night.

Although *Resolution* was out of sight, she was never out of mind. John and I had discussed much of what had yet to be done, and my head swirled with visions of chain pipes, foredecks, engine connections, thru-hulls, bilge pumps, and bronze keel bolts. I didn't sleep very well on the extra-firm mattress in the hotel, and was up long before the wakeup call. The flight was off the ground at 8:55 and landed at Narita just before noon. The girl stood just outside the exit doors when I left the customs area. T'was good to be home after months of drilling and fastening on *Resolution*.

All three grandchildren and all four daughters awaited me at home. If you think drilling holes in concrete for 9mm bolts causes a ruckus, try three grandkids all mad at the same time. Still, with sushi from my favorite restaurant and the family gathered 'round, my first evening back in Japan was satisfactory, if not quiet.

I spent a good part of Friday at *Kazi*, Japan's biggest yachting and motorboating magazine. Ken Ando, one of the editors, interviewed me for an article in the next issue, and we selected photos and cleared up details about an open boat cruising article I'd written in Japanese for the magazine.

Saturday saw me on the road in my Isuzu SUV, headed for the Yokohama Yacht Club (YYC), my boating headquarters in Japan back then. These days, my Blue Water 21 *Endeavor* resides at Funabashi Boat Park. Minoru Saito, seven-time round-the-world single-hander and the oldest man to sail alone around the globe without stopping, moored his boat, *ShutenDohji II*, at YYC. He'd dismasted her on his way to Guam and was re-stepping the mast after sawing two feet off the bottom. "Gotta go to Brisbane," he said. "Only place I can get a new mast."

Then I visited Kodansha International, which published both my *Seeing Japan* and *Inspired Shapes*, a book I translated from Japanese to English. Kodansha, the parent company, was interested in a Japanese-language account of my single-handed voyage around the world.

Now, nearly a decade later, I read these words and wonder what my life would have been like if things had gone according to plan.

Still, I couldn't forget *Resolution*, and soon it was time to climb on a Korean Airlines plane for Auckland via Seoul. But I'd seen my children, their spouses, my grandchildren, the girl, and Tomlin the dog. So I could go back to building my ship of dreams.

In New Zealand, I built *Resolution* step by step, and completion of many small operations would net me an ocean-going boat in the end. Of course I counted the days until *Resolution* would be launched and we could begin her shakedown cruises, but as usual, my estimates were far off.

Helping Hands Always Welcome

Outside the boat, I worked on the massive rudder, fitting and gluing its hardwood backbone to the kauri blade. Once they were glued together, I drilled through the blade with an extra-long 6mm bit, tapped epoxy-coated threaded stainless steel rods through the holes, and lugged the nuts down on washers fore and aft. A blade of two layers of laminated 9mm plywood went on the bottom of the rudder, attached with hefty doses of thickened epoxy and 3" bronze screws. By the time I finished, little could cause that heavy rudder blade to break. And in the end, it never did.

My friend Blair Cliffe, whom unfortunately I missed during my 2018 trip to New Zealand, came to help for a couple of days. He held a first mate's ticket and his boat supplied the offshore gas rigs. Blair spent two weeks on board and two off, and he sneaked away from his two daughters in New Plymouth to come help build *Resolution*.

He arrived on Thursday, shortly after nine. I knew he was coming, and had some jobs laid out. His Volkswagen bus, of indeterminate but vintage age, readied him to camp out at the club for the night. "Pull on your coveralls," I said. "You get to spread epoxy."

Moments later, he'd changed into a pair of faded red coveralls. "What do you want me to do?" he asked.

"We'll put a transom on the boat," I said.

"Just show me how." Blair was more than ready to go.

I'd prepared a 12mm blank for the transom, but we needed to epoxy a 6mm sheet of ply over the top to bring the blank up to proper hull thickness. I pulled a 4x8 sheet of 6mm ply from the stack of plywood standing against the wall. We laid it on the building jig, put the 12mm transom down over it and traced the outline. A couple of minutes with a Japanese handsaw gave us a 6mm panel the same shape as the fiberglassed 12mm blank. Now to epoxy the two panels and stick them together.

Good job for a novice like Blair. He pulled on a pair of surgical gloves, and went to work learning how to mix the epoxy (no big deal, the pumps meter the right amounts of goop into the container). Mixing in the microfiber took some getting used to, but he made it. With one side of the 6mm sheet painted in epoxy and the 12mm blank covered with rows of epoxy spread with a serrated trowel, all we had to do was lay the 6mm piece on top of the 12mm one and place chunks of cut stone 75mm thick on top to press the two pieces firmly together.

Friday morning, I took Blair to breakfast at Fran's. He chose eggs Benedict and I had an omelet. The café was crowded with Cambridgeites and pleasant greetings filled the air. Lovely Lacy, Fran's beauty queen waitress, spread her smile across the room and Fran's good food disappeared with record speed as her customers started the day. Satiated with Fran's fine cooking, Blair and I drove a quarter of an hour to the workshop, where *Resolution*'s transom waited beneath a layer of stones.

We found the 6mm flat firmly bonded to the 12mm blank. After a bit of trimming, we were ready to start applying the transom to *Resolution*'s stern. Blair hoisted the thick plywood into place and I clamped it to the quarterdeck beam. We penciled in the cockpit drains, took down the transom, and cut the drain holes out with a cordless drill and a saber saw. We reapplied the transom to its proper place on the stern. We marked all the framing where bronze screws would hold the transom in place. We trimmed the drain holes to match the ones running through the smelly locker. Then we lowered the transom to the floor, laid it on the jig, drilled all the screw holes through from the back side, turned it over, and bored countersinks into every hole.

"So, are we ready to put that transom on?" Blair seemed anxious. He had only until early afternoon so we needed to hustle right along if we were to get the transom hung.

"One more step," I said. "Help me lift it up into place once more."

We lifted it and clamped it to the beam. Then adjusted all around so the transom was in just the right place.

"Okay," I said to Blair. "Now you can drill the screw holes in the framing. Just take a 3mm bit, make sure it's shorter than the bronze screws, and drill a pilot hole through each of the holes in the transom."

Blair did exactly that.

Off came the transom again. "Epoxy now?" Blair asked.

"One more job."

"What's that?"

"We need to take the vibrating sander with 120-grit paper and rough all the framing surfaces."

"Right."

I got the sander ready, and Blair did the sanding. "I've never built a boat before," he said. "Want all the experience I can get."

"Now you can mix up a pot of epoxy and microfibers," I said. "Make it about the consistency of mayonnaise."

Blair's grin was a mile wide. He'd been waiting all this time to put a major component on *Resolution*.

Before he put epoxy on the transom, I drove two bronze screws through the upper corners so that they stuck out about 3mm. "Okay, smear epoxy between the lines marking the framing. I'll put goop on the framing itself."

"Right-o," he said, and proceeded to slop thickened adhesive onto the transom.

Once everything was epoxied, we lifted the transom into place, made sure the two screws I had started seated into the correct pilot holes, then drove the screws home with my Black & Decker cordless drill. That secured the transom into place, and all we had to do was put a total of 93 bronze screws into the holes we had prepared for them and tighten them down. Of course, epoxy squeezed up from the framing, so Blair got to run his vinyl gloved finger along the visible frames to create a fillet of epoxy between transom and frame.

"Wow, looks good," he said, more than a little pride in his voice.

"Does, doesn't it?" Pride in my voice, too.

So you see, some days all you accomplish is measuring, cutting, and installing a small panel for something like a tool locker. On other days, something momentous happens, like hanging the transom.

Resolution, *Continued*

Work on the cabin and the cockpit proceeded at the same time. It's not like you can just pick a part of the boat to finish off, then go on to the next part. Many of the parts interlock, so work on the cabin and work on the cockpit must go hand in glove. With Blair's help, the transom was now in place, so work could continue on the cockpit, the smelly locker, the long lockers, the seat backs, and the aft face of the cabin, which came down to the bridge deck beam.

Resolution had a problem with her bridge deck beams. Whether the problem was one of my making or one on the plans themselves, I didn't know, but the top of the bridge-deck beam on the No.8 bulkhead came up just to the bottom of the bridge-deck beam on the No.7 bulkhead, which was also the beam that supported the aft cabin bulkhead.

To deal with the differences, I epoxied flanges of 9mm plywood to the undersides of the No.7 bulkhead to give the bridge deck a land about 30mm wide. I made sure the screws went into the longitudinal cockpit framing at about 100mm intervals. The bend of the bridge deck forced the plywood down against the flanges and well slathered with thickened epoxy, the seam was very strong.

In a couple of days, I installed the cockpit floor, which is 12mm plywood, the bridge deck, the cockpit locker floors, and the cockpit sides. Time to paint.

Winter

Of a winter's morning, I would drive 16 km from the club in Matangi to Fran's in Cambridge. Frost blanketed the lawn at the club and the maples along the road dropped their fiery leaves, one by one. Paddocks lined State Highway 1B all the way to the dairy products plant on Victoria Road. Black and white Holsteins steamed in the morning cold. Sheep huddled beneath spreading branches, wearing next Christmas's ugly sweaters. A farmer and his wife guided a herd of cows across the road. Drivers waited. Dairy products are New Zealand's biggest export commodities. You give way to cattle.

My second winter in New Zealand, and my alarm screeched in the half-light of dawn. The hands read 6:30. Time to rise, gobble down my morning ration of All-Bran, banana, and soy milk. I left the club a few minutes after seven, still driving with my lights on, headed for Fran's Café and an hour of writing over hot lemongrass or chamomile tea.

Often as not, Highway 1B was cloaked in fog. The black tarmac stretched forward into nothing. Pinpoints of light became onrushing cars when they were a hundred meters away. Usually the fog burned off by the time I left Fran's at quarter to nine.

Of a winter's night, I sat in my little trailer with the small heater on high. Outside, a smattering of rain. The radio said the temperature was 7 degrees Celsius. I brewed a pot of herbal tea, a piquant mixture of lemon and ginger, and settled down to read a book. Food for the brain.

A Warm Place Out of the Weather

For months, the cabin gaped, a huge hole in the middle of the boat. First, I built the foredeck and then the cockpit, seats, and seatbacks. Even the transom got hung before I started work on the cabin.

The first indication that *Resolution* would even have a cabin was the aft bulkhead—12mm plywood, cut to shape, then glued and screwed to the No.7 bulkhead's bridge deck beam.

I moved forward.

Using the after end of the foredeck as a form, I laminated a hardwood beam—eight layers of 8x50mm strips, clamping them to the edge of the foredeck. When the epoxy had cured, I took the beam off (no epoxy between the first strip and the deck), planed it down to 45mm, and bolted it back onto the deck in a bed of epoxy. Six bolts in all. Nothing was going to move that beam!

The cabin's 12mm forward bulkhead, fit on the forward side of that beam, glued and screwed. Thirty centimeters forward of the forward bulkhead came a false bulkhead. Not false in that it's weak or insignificant, just that it can't be seen from the inside of the cabin. The false bulkhead came directly above the No.4 frame, so it sits on a beam. I laminated two layers of 20x45mm kauri to the foredeck over the beam, and attached the false bulkhead to that, glued and screwed to the forward face.

The false bulkhead forms the forward end of the cabin on the outside. Between the false bulkhead and the forward cabin bulkhead reside two big deck lockers, port and starboard, and two dorade vents, inboard.

Next step: 45x45mm corner posts at the inner edges of each bulkhead, shaped to fit at the carline—the beam that supports the side deck—and at the roof point.

At the same time, I laminated inner beams for the bulkhead rooflines. Way back at the beginning, I laminated a No.6 beam for the cabin roof. That beam served as the form for the inner beams. I laminated 50x8mm kauri strips to the inside of the beam to make a second beam eight strips thick. Then I ripped that beam down the middle and planed the two resulting beams down to 20mm each. These inner beams reinforce the rooflines of the after and forward bulkheads.

With corner posts and inner beams in place, I installed 25x45mm rails fore and aft, forming the roofline of the cabin sides.

Next came two posts at the No.6 frame and the cabin roof beam, with hanging knees. The posts were 60mm laminates of three lengths of kauri. The beam was five laminations of 8mm kauri. And the knees were eight lengths of kwila hardwood laminated in a curve. Trimmed down and shaped, they're glued to corner pieces to complete the hanging knees. Then the knees were glued and screwed to the posts and beam with three-inch bronze screws from the inside and two-inch stainless screws down through the top of the beam.

After I put gutter pieces at the sides, the 9mm side panels could go on. The gutters were of 20x60mm kauri, shaped on the router to catch the condensation off the cabin walls and trimmed to fit the carlines. I glued them to the carlines, leaving a 20mm land for the cabin sides.

The 9mm side panels started as a rough-cut piece 40cm wide and the length of a sheet of plywood. This I trimmed to the curve of the deck. Then I ran a pencil around the inside of the frames to mark where to glue. Epoxy on the framework. More epoxy between the pencil marks. Stand the ply against the frames. Screw in place with stainless woodscrews. Layer one, done.

The second layer of 9mm gets glued and screwed over the first, so the walls of *Resolution*'s cabin are a solid 18mm (11/16") thick. You thump them, they ring solid.

John designed a cabin roof of tongue and groove strips. But after losing a finger and part of a thumb to the bench saw, I was a bit leery of the beast and decided to do the roof the same as the decks—two layers of 6mm ply over stringers.

I used four sheets of 6mm ply, two per side, to give *Resolution* a solid cabin roof. I had to trim, plane, fill screw holes, sand, and cover the entire roof with a layer of 10 oz. biaxial woven roving in epoxy. A labor of love.

You may wonder why I included all the detail about building the cabin. Two reasons. Having read this far, you know how a lot of small jobs come together to create a large result. Further, someone who's actually building a Sundowner may find this explanation helpful.

Quarterdeck in the Stern

Thirty-six centimeters forward of the transom, I fit the five-layer quarterdeck beam. In the center, I notched in a crosspiece of kwila hardwood, and another of kauri on either side. Planed and sanded, the support structure was ready for its double-layer skin of 6mm marine ply. I used offcuts from the cabin roof to plank the quarterdeck. Two days and the decking was glued and screwed (I love that turn of phrase). I trimmed the forward edge in a pleasing curve that led into the trailing edges of the coamings.

Two 65mm square mooring bits, named Delilah posts by their designer, protrude from the decking near the forward edge. They're screwed to a 12mm plywood gusset that looks like a hanging knee beneath the quarterdeck. Solid? You could attach a rope and lift the entire boat with one of those Delilah posts.

Wintertime at 37 degrees south means the sun comes up after 7 a.m. and sets before 5:30 p.m. There was not a huge daylight window to work in, and I finished many a task under the three roof lights in John's workshop. Work progressed, slowly, but *Resolution* neared her final stages.

A lot of people give up just before they're about to make it.
--Chuck Norris

Chapter Eleven

Maybe We're All Dreamers

    Resolution had a cabin top. And her cockpit was nearly finished. I could sit in her well-protected cockpit and imagine how it will be out on the ocean. I came up with alternative lyrics to John Lennon's "Imagine."

> *Imagine open ocean*
> *Sails drawing full and by*
> *No rocks below us*
> *Above us only sky*
> *Imagine you are sailing*
> *If only for a day...*
>
> *Imagine all the countries*
> *You'll be sailing to*
> *Meeting lovely people*
> *Like other cruisers do*
> *Imagine that great feeling*
> *An inner life of peace...*
>
> *Maybe I'm just a dreamer*
> *But I'm not the only one*
> *I hope someday you'll join me*
> *And sail the world alone*

    Resolution felt like a boat. I sat on the quarterdeck and surveyed my domain. In my mind's eye, she leaned into the waves, main and jibs drawing, with the wind abaft the beam. Maybe she'd be making as much as five knots. But she was still in her cradle, and had no lead ballast in her keel.

In the *Sundowner Diaries*, John describes the situation like this:

"I was very pleased to be able to take a prospective Sundowner builder up there and sit him comfortably in the deep and sheltered cockpit while he looked around, eyes wide at the amount of space and comfort in this little cruiser, only about two feet longer and a foot wider than his own trailerable yacht. He climbed around down below, asking why and how about key features, moving about to get a feel for the space and I can imagine that he will be seeing the same sort of images of cruising destinations in his mind's eye that I do when sitting in there looking out with unfocused eyes and wandering mind.

"She's coming along. Charlie is busy making the patterns that will go off to the foundry to be pressed into the sand, making the molds into which the molten bronze will be poured to create the rough pieces that after a lot of filing, drilling, and grinding will become finished fittings ready for the boat.

"Around the cockpit area he has completed the quarterdeck and the area around the stern end of the cockpit. There are two mini Samson posts there. We have jokingly called them "Delilah posts" as we have a bigger Samson post up in the bow and we have all three so well braced that any one of them could support the entire weight of the boat. These after ones are not only there to provide tie-up points for the mooring lines but they also have to cope with the stresses of lying to a parachute anchor in a serious storm, or dragging warps, so they need to be tough. That calls for kwila, one of the strongest of woods, and able to cope with harsh environments without paint or varnish to protect it.

"That quarterdeck has a hard point built in that will be drilled and fitted with a hardwood pad to hold a vise when needed. It's at the right height to be able to attack a metal bar with a hacksaw, to splice a piece of wire rope, or to hold a piece while shaping it with a file or chisel. A good workshop should have a vise, solidly mounted and accessible and a small boat that is going to be going where Charlie is planning to take *Resolution* really needs a good workshop. It's a long way to go to find someone who can fix it for you so it's all DIY and the cockpit will be a maintenance center, a woodworking shop, a rigging shop, and an engineering works as well. So, a vise, a hardwood pad where things can be hit with big hammers, where the skipper can swing a spanner or big wrench, or chop with a chisel, is not only useful, it's essential.

"Much of this boat is two laminated layers of wood; the hull itself, cabin sides, cabin top, decks and quarterdeck are all done this way. As each part of the boat gets fitted and glued up, she gets more and more rigid. Now that the 'lid' is on, she feels incredibly strong, reassuringly so, and it's a feeling that's very comforting when the immensity of the planned voyage is considered. Hey, it's big water out there and deep in the Southern Ocean— about halfway from New Zealand to Cape Horn is the spot which is the most distant from land that it's possible to get on this planet."

As John said, I'd spent two weeks making patterns for the bronze fittings to go on my *Resolution*. Originally, when John first drew the plans for the hardware, he thought he'd found a professional pattern maker, but things didn't work out. I had a stateside foundry make six bronze stanchion bases from patterns they had in stock. By the time they reached New Zealand, they'd cost me close to US$1,000, including New Zealand's healthy 12.5 percent general sales tax. Imagine the price for all the bronze *Resolution* needed.

Luckily, John remembered a one-man foundry in Tauranga, which he said would cast the hardware at a ridiculously low price—which was not exactly true. Still, Roger Woodhouse, proprietor of the foundry, went out of his way to bring the hardware in at a reasonable cost to me.

By the time I got through making wooden patterns for cranse irons, chain plates, bow rollers, stem pieces, whisker stay and bobstay plates, Samson post cross pieces, two sizes of pad eyes, backing plates, rudder gudgeons and pintles, and so on, I had a total of 50 pieces of hardware to be cast. At first Roger estimated 90kg of phosphor bronze at a total of more than $4,200.

I screamed—silently, of course—and said it was too much.

Then Roger found that he could get silicon bronze for less and that it would take less to produce the 50 items as well, 80kg in all. Then he figured a way to make the molds more efficiently, and gave me a final quote of just over NZ$3,200, including the government's take. A big bite from the budget, but a grand less than before. I gave him the go-ahead.

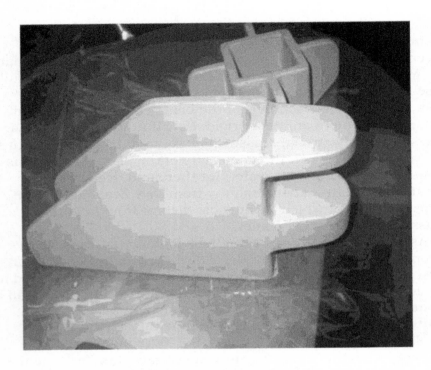

Although the foundry was more than 100km from John's shop in Hamilton, we asked Roger if we could come over and watch him cast the hardware. He agreed, and 10 days after I delivered the patterns, Roger fired up his supercharged oil furnace, lowered his ceramic crucible into the whirling flame, and started melting the silicon bronze ingots he'd gotten from Auckland.

John and I left Hamilton about 9 a.m. as we had two places to go before we hit Silverstream Foundry. Again, I'll dip into the *Sundowner Diaries* on jwboatdesigns.co.nz for John's description of the pour:

"We have just had a really good day watching foundryman Roger Woodhouse making *Resolution*'s cast bronze fittings. Charlie had spent a couple of weeks making the patterns from my drawings, using MDF sheet, five-minute epoxy and car body filler. It took a while but he made a light wooden (sic) replica of each item needed. We took those over to see Roger to make sure that they could be used for making a mold, and to get an estimate of the costs involved.

"The first estimate was a shock, a lot more money than we thought, but some negotiation and a change to a less expensive metal brought that back to a workable level, and Roger issued an invitation to come over and watch as he did some of the pour.

We arrived to find the furnace roaring away and several bars of what looked like pale gold ready to feed into the crucible, a double row of two layer steel boxes full of black sand lined up on the floor, and Roger busy pressing the stemhead fitting pattern into another box. Some of the patterns are quite challenging from a foundryman's point of view—thin sections connecting with thick ones, tight radius intersections, and deep thin sections that could only be set edge on into the mold.

"Roger's great; he'd thought it over and figured out how to do it. My homework (read several books that I found in the library) seemed to have been helpful as no modifications to patterns were required, plus a large dose of his skills made for a successful day.

"It was quite spectacular to watch. The roar, the heat, and the glow radiating from the furnace are really something even in this small operation – the careful handling of the "pot" full of glowing hot metal that is so liquid it flows like water, the heat radiating off it is such that you could grill your steak from several paces away, and the whole thing looks like the inspiration for some of the underground volcanic scenes from 'Lord of the Rings.'

"Roger and his helper wheeled the crucible over to the molds and literally just poured it in. This is metal, tough metal, and watching it being poured in like water from a tap is quite something. We stood well away, just in case, as our light shoes were not the thing to cope with a spill of molten metal.

"A while later a mold was picked up and split, shaken out, and the castings pulled free from the sand. Even sitting on the concrete floor they had an air of solidity about them that the wooden patterns did not have. I can tell you that if you are building one of these, then there is no need to make the patterns any heavier than I have drawn them. These are strong, they have quite a reserve, and while I had calculated the strengths, it's quite different to see the real thing.

"Another stage of the build done. There is a lot of cleaning up, drilling and countersinking and polishing to do, but it's more progress.

"Fun? You bet, it feels quite primal, casting metals."

At the foundry, Roger trimmed and fettled the pieces, then tumbled them in sand. On a Wednesday, nearly six weeks after I started on the first pattern, I picked up three boxes of bronze hardware and a box of MFD patterns. The only thing left to do was drill holes.

No 21-footer has ever been equipped with the kind of massive hardware that my *Resolution* had. This boat is truly "exploration grade."

While Roger was setting up to pour *Resolution*'s hardware, I cut a big hole in the foredeck—space in which to place the fore hatch.

When John went on vacation to the United States in September 2005, one of the tasks he left for me to do was to prefabricate the fore hatch. The process included laminating a curved hatch roof and cutting inner coamings and exterior hardwood hatch sides, along with some rather intricate corner pieces. I did the job as well as I could. And wonder of wonders, the prefabricated pieces fit . . . well, mostly.

First, I glued the inner coaming together using corner clamps and epoxy, lugged it to the foredeck, and trimmed the hole until the glued and screwed box fit snugly inside. I slathered thickened epoxy on the forward deck beam, the stringers (John calls them Princess planks) port and starboard, and on the King plank at the after end. I slipped the coaming box into the opening, tapped it carefully into place, then screwed it to the beam, stringers, and King plank with 40mm stainless screws. Solid.

Fit and attach the corner pieces. Lay 15x20mm battens along the deck next to the coaming and screw them tight. Then fit tigerwood outside panels and the deckside portion of the hatch is almost finished.

Why?

Why leave, they ask, the ancient island empire of *Zipang*, where supple geisha flow through timeless *jiutamai* in time to twangs from catgut *shamisen* strings and hollow echoes of bamboo *shakuhachi* flutes, where cherry blossoms reach full bloom and fall as a young samurai in battle, where incense and the chant of Zen from temples unchanged for millennia herald morning meditation, where the concept of face still means a father is responsible for what his children do? Then the Little River Band strikes a chord when it sings, *if there's one thing in my life that's missing it's the time I spend alone, sailing on the cool and bright clear water.*

Three decades I spent in the shadow of the emperor's palace, where lights never dim and commuter trains run far past midnight so drunken businessmen can find their way home for a few hours before taking up the corporate banner with the coming of the sun and riding the same train back to work in clouds of alcoholic haze. Then I lifted myself by my bootstraps and moved myself south, far south, until I was under the Southern Cross and hoped it would bring me a brand new day.

Back then, back in previous days, I had another boat. Her name was *Umisaurus*, a big center-cockpit sloop. I often took her south to Oshima Island, like as not with half a dozen scuba divers aboard. Sailing through the night from Tokyo Bay put us midway to Oshima as the sun rose just after 4 a.m. A band of coral would spread across the horizon, a patch of the Pacific so wide it looked like you could see the curvature of the Earth. The Pacific's long groundswell and the peeking sun striped the water with a line of gold. I often caught the *chuff* of dolphin breath even before they appeared. There, flinging drops of gold from their dorsal fins, a half dozen bottlenoses. For a few moments, they'd cavort in the bow wave while the divers oohed and aahed. After dolphins visited, I always got the feeling it would be a good day. But then, any day out on the ocean is a good one.

Back then, the girl waited, though often visited by her grandchildren. I'm sure she wondered why I worked alone spiling planks and setting Samson posts and mounting rub rails and casting fairleads and stem heads and pintles and gudgeons for my *Resolution*. Soon my boat was to leave her cradle for the rippled harbor at Tauranga and the fabled Bay of Plenty, from whence we'll set off to sail 20,000 miles around the globe along the 40thparallel; just me and *Resolution*. So why did you do it? Asked, "Why climb Everest?" George Mallory told a newspaperman, "Because it's there." Mallory perished on his third attempt to scale Everest in 1924. No one knows whether he made it to the top or died trying. But there's your answer to the why of sailing alone around the world. *Because it's there my son because it's there, and you must try the task to see if you can, to see if you can.*

The Plans of Mice and Men

There I was, into the third year of building *Resolution*. When I landed in Auckland on August 30th, 2005, I fully expected to have *Resolution* in the water by November 2006. Never did John Welsford disabuse me of that idea. In fact, he never, ever, said a word about my progress, or lack thereof, in building *Resolution*. He's never been on my back to get things done; he never pointed out how slow I am at times. But he did start making noises about building his own boat, "as soon as my shop is free," he'd say pointedly, a big grin on his face. But, like me, John doesn't always get things done as quickly as he originally plans to have them done. Surprise, surprise. Perhaps we're both human.

Ship of Dreams

The departure date for the voyage around the globe kept getting pushed back. January, I said then. That will put me at the Horn in March, which by all accounts is as quiet a time to sail in those high latitudes as there is. I knew from sailing in Canada in September that nights will be very very cold, and I planned my dress code. In the straits of San Juan de Fuca, I layered with cotton underwear, sweats, and foul weather gear. Not enough. For the circumnavigation, I planned full body under-layer of polyester or the like, a full body layer of fleece, and then my foul weather gear. Same with my feet. Tight socks of polyester, second layer of New Zealand possum and wool, sea boots. John even gave me a pair of waterproof mittens to wear over my cold-weather sailing gloves. I looked for anyone with advice on how to keep warm, say in Alaska. I wanted to hear it all before I left.

We Dream of Voyages

Let Me Tell You a True Tale of Crossing the Bar.

Crossing the Bar

Sunset and evening star,
And one clear call for me!
And may there be no moaning of the bar,
When I put out to sea,
But such a tide as moving seems asleep,

Too full for sound or foam,
When that which drew from out the boundless deep
Turns again home.
Twilight and evening bell,
And after that the dark!
And may there be no sadness of farewell;
When I embark;
For tho' from out our bourne of Time and Place
The flood may bear me far,
I hope to see my pilot face to face
When I have crossed the bar.

--*Alfred Lord Tennyson*

We didn't wake until seven, so by the time we'd eaten breakfast, the clock had chimed eight. Humberto cast off with me on the tiller and Yoshi tending the starboard line. The diesel rumbled pleasantly below the cockpit floor as we sidled up to the fuel dock of the Port Angeles yacht harbor. We took on 14 gallons of diesel to feed the engine, topping off the 68-gallon tanks.

We left the harbor, turned north, and then west toward the open ocean. Tonight there would be no quiet berth in a friendly marina, only the long Pacific swell and the 10-foot chop that would meet us at the bar.

I'd never been across the bar, but I knew it was called Juan de Fuca's Graveyard. And I knew history recorded 137 ships sunk at the bar between 1830 and 1925. Seated in the cockpit of a small sailboat with one eye on the horizon and one on the compass, I found it less than comfortable to contemplate the 250 souls who perished when the sidewheeler *Pacific* went down on the bar, south of Tatoosh Island. We drew six feet of water and should be able to safely cross over the same route that sent the *Pacific* down.

Should.

I checked the GPS and plotted our position on the chart, then changed our course to 265 degrees. My watch read 16:04 and our ETA for the bar was nine at night.

The *Susan Abigail* is counted among the ships that died in the Graveyard. But she was actually murdered--captured and burned in July of 1865 by the *Shenandoah*, a Confederate States of America raider that made a career of sinking U.S. whalers, even after Lee surrendered at Appomattox Court House.

The *Shenandoah* was launched in Scotland as the *Sea King*, a troop transport, in 1864. The CSA bought her, christened her *Shenandoah*, and fitted her with eight cannon—four 8-pounders, two 32-pounders, and two 12-pounders. Under the command of Lieutenant J.I. Waddell, she sailed for Australia, taking five prizes along the way. From Melbourne, she moved against the American whaling fleet, and sank at least five, burning them to the waterline. Lt. Waddell and his raider arrived in the Northwest in late summer. He learned the war was over on August 2nd, 1865.

Some say Waddell then camouflaged his raider as a merchant ship and returned to Liverpool. But the natives of Nitinat Point say an iron gunboat lies buried in the sand off the entrance to the Narrows, and many believe it is the *Shenandoah*, sent to her fate for the 38 ships she sank.

At 17:45, our position is 48 degrees 20' North and 124 degrees 23.5' West. We decide against the southern route and set our course to skirt the navigational buoy to the north of Tatoosh Island. Our five-knot speed should put us over the bar at about 20:00.

The seas begin to build at 19:40. The distance between waves shortens, indicating shallower waters. Ahead in the moonlight, I see a dark patch spreading toward westward. The waves no longer show any pattern—thrown upward by the shallow bottom, thrust eastward by the Japan Current, blown southward by the northerly wind—confused and choppy, at least 10 feet, maybe 12 from trough to crest. I strike the main and drop the staysail.

We motored through the chop. Our boat *DoriKam*, a Westsail 32, took the waves with her buxom bow, throwing spray out and away from the boat. I try to steer into the waves but their confusion thwarts me. So I concentrate on maintaining the course at 275 degrees.

The dependable diesel never misses a beat. It swings a 14-inch propeller that gradually pushes us past the navigational buoy. Fifty yards. A hundred.

To the south, the water glistens in smooth patterns. I push the tiller to the starboard. The boat noses around toward the south. With my eye on the compass rose, I put the tiller amidships when the bearing line reaches 180 degrees.

The dark mass of Tatoosh Island passes to port. I see the Hole in the Wall, with the fangs of Jones Rock protruding from the tides. I set the autohelm and clamber out on deck to hoist the main and set the staysail and the yankee. The breeze comes off the Pacific at about 10 knots. The cutter gurgles and settles down to her south-bound heading. The boom jumps with the motion of the boat, so I vang it to a portside stanchion.

The wind softens. *DoriKam* slacks off to 4.2 knots.

At 02:00, we alter course to 170 degrees, a dozen miles off the coast of Washington state. The sun clears the coastal mountains at 06:37. We've crossed the bar and *DoriKam* is not among the vessels listed in Juan de Fuca's Graveyard. Would that the same fate had awaited *Resolution*.

Ask not what tomorrow may bring, but count a blessing every day that fate allows you.
  --Horace

Chapter Twelve

Keeping the Dream Alive

In October 2007, I wrote:
Time draws near. In December, *Resolution* should kiss the sea at Tauranga. Then, after six weeks to two months of shakedown cruising in New Zealand waters, she and I will turn bows east and sail for our first port of call, Stanley, the Falkland Islands.

We'll sail as summer wanes, Indian summer arrives, and near Cape Horn, autumn calms arrive. Pilot charts tell me the weather in March near the Horn is about the best of the year. At the end of March, of course, we'll have 12 hours of daylight and 12 of night. And experience in Canada in September tells me days will be lukewarm, nights downright cold. That's how it will be in March in the Southern Hemisphere.

Nevertheless, *Resolution* and I are determined to go. We'll stop in Stanley, shop at the local grocery store, and push off north toward warmer latitudes on our way to Hout Bay in South Africa.

My Japanese friends, Minoru Saito and Ken'ichi Horie, have sailed these oceans several times. In fact, 73-year-old Minoru now plans a west-around solo circumnavigation. He bought a 56-foot steel ketch in Hawaii last month and he'll bring it to New Zealand for refitting in December 2007. He plans to leave Japan in October 2008 on his westabout voyage, so I doubt that our paths will cross, except here in New Zealand before I leave.

Charles T. Whipple

How to Win Friends

One of my current goals is to walk 12,000 steps per day, or more. It takes me about two hours, slow as I am.

When I first started toward this goal, I was lucky to do 3,000 steps, but soon I was walking the paved walk along the drainage canal that runs from about 20 meters west of my house all the way to Tokyo Bay. It's slightly over 3,000 steps from my front door to the overpass of the Keisei Railway. There, in the warmth of the morning sun, turtles bask on the rocks at low tide. At high tide, varicolored *koi* carp swim in languid rhythm upstream and down, over and across. I wonder who turned these *koi* loose in the canal. A purebred one could run US$10,000. A white egret often stands on one leg, keeping an eye on the slow current of the canal. What it hunts for, I have no idea. Its patience is edifying. Would that I had as much.

Nowadays, during the day, I drive to Aeon Mall near Chiba's Makuhari area and walk up and down in front of the shops. Most people recognize me now. Often I stop at Starbucks for a chamomile tea and sometimes a chocolate scone (I acquired my taste for scones in New Zealand). I write or rewrite or revise there at my regular table until the chamomile is gone. Actually, I've written some 25 novels, most of them at Starbucks or Mr. Donut. For some reason, the creative juices flow more smoothly there than among the clutter in the cave that serves as my home office.

I walked a decade ago in New Zealand, too. And I walked during my re-visit, too, and I had no need to even prompt John to do so. You see, Indy the truck guard dog also likes to walk. One of his preferred walking routes is the seriously long beach at Orewa. A very good hike for an old duffer with arrhythmia, but we made it.

As I said, the beach at Orewa is very long and very smooth. It's a shallow one that extends hundreds of meters off shore before getting over waist deep at low tide.

Families walk the beach at Orewa. Couples walk the beach at Orewa. Friends walk the beach at Orewa. It's open and appealing and you don't have to leash your dog if he's smart enough to come when called. Indy's smart.

Ship of Dreams

When I lived at WOS, I walked. Sometimes in the evening, but always the same route. Once my Kiwi physician sent me to an x-ray clinic, as I remember. I had an appointment and walked up to the reception desk as if I were on an important mission.

"Hey! I know you," the woman at the desk said. "You're the bloke who's always walking over by Matangi."

"I am," I said, flattered to be recognized by a total stranger. But that's kind of the way it is in Kiwiland. One moment you're strangers, the next you're good mates.

Shutting out the Weather

Before, I wrote about cutting a hole in the foredeck for the forward hatch. Well, the hole was filled with a tight-fitting square of Fijian kauri. The square was outlined on the outer deck edge with battens 15mm wide and 20mm high. They ran from corner piece to corner piece and supported both the deck edge and the box, which formed the hatch's coaming. Outer coamings encircled the battens (although they, too, are square) and butted up against the outer corner pieces. Holes drilled through those outer coamings just above batten height (20mm) let any water that makes it past them run right back out onto the deck. Helps keep the forward cabin dry. And 6mm bolts go down at each outboard corner, tying the outer coaming solidly to the Princess planks (John's term for stringers that are larger than ordinary ones but smaller than King planks).

When John went on vacation that time it rained day in and day out, one of the jobs he left me to do was laminate two hatch covers from 4mm plywood. Did that. Cut all the lumber pieces, too (and they fit). Those hatch covers gathered dust for nearly two years before the smaller became the lid of the forward hatch. John designed that forward hatch to open from aft, expecting the lid to lie flat on the foredeck. Hah. Wouldn't work. First of all, the Samson post is in the way; second of all, the anchor chain winch will be in the way, too. Quick reversal. Open the forward hatch from the forward end. That meant fashioning hinge blocks from kwila and screwing them in place at the outboard edges. Still, all in all, it came out well.

With the forward hatch done, hinges and handle attached, I filled and sanded the foredeck. And even though I still had to glass the forward and aft ends of the deck house, I felt I should get on with the main hatch. Again, the main components were made nearly two years ago. Unlike with the forward hatch, the rails for the main hatch were too high. I lopped 50mm off the bottom, on a slant, of course, so the forward end would be high enough to accept the hatch cover runners.

I made a lot of adjustments in the main hatch. John designed it 800mm wide. I reduced the width to 700mm, which was just enough for me to sit in the companionway with my elbows up on the hatch slides. It was also the width that resulted when I screwed the hatch's inner coaming pieces to the stringers in the cabin roof.

John also designed the main hatch 800mm deep, but I chopped it to 750mm, which seemed right. That allowed the hatch cover to open to a 700x700mm space above the kwila companionway steps.

The inner coamings led over the top of the cabin forward, serving as hatch cover slide rails. Outside these coamings came the main rails of tiger wood, a VERY slick hardwood. I'd cut grooves with the router. With the rails in place, I could build the hatch cover. First I installed retaining strips port and starboard. They were milled hardwood pieces with stainless slides fitting into the grooves. I held them in place with small pieces of 6mm plywood screwed to the inner coamings. This kept the runners solidly in place. Thickened epoxy spread on the tops of the runner pieces readied them for the hatch cover, which I carefully placed atop them, and screwed them tight. A day later, with the epoxy hardened, I removed the plastic barrier sheet and the hatch cover slid cleanly back and forth on the rails, slick as a whistle.

Not yet complete, the hatch cover sat on the workshop floor. A laminated beam glued to the inner aft end would serve as abutment for the washboards. Another laminated beam at the forward end formed the seat for the end piece. Once it had a handle, the main hatch cover was complete, except for a layer of fiberglass cloth yet to be epoxied across the top.

Outdoors at Last

Murphy's Law is not absolute. Some things go well.
"Ready?" John said. "Let's move her outside."

If you remember, John, Andrew, and I turned *Resolution* in August 2006 with two four-part tackles and two lengths of galvanized pipe. The cradle I made for the turnover supported *Resolution* for months while I got her ready to move outside. Now, with the topsides protected by a coat of primer, she could stand the outdoor UV rays.

Two friends from Japan, Masaya Kinoshita and Stanley Fong, came to visit just in time to assist with the move.

The first move involved John's new 2.5-ton hydraulic jack. Masaya pumped up the bow end, then the stern, while I slipped 2x12s beneath the cradle and nailed them in place; then rearranged the galvanized pipes to give the boat its rollers. Stanley watched.

The boat didn't face directly out the big double doors, so it had to be slewed around to starboard so the port side would clear the doors. Then we hooked a four-part pulley to the front leg of the cradle, and John and I gave a big heave. *Resolution* swung around like a well-trained horse.

As soon as we were certain the boat would clear the doors, John backed his Camry to where its trailer hitch could be used as the anchor point for that four-part pulley, and we started heave-ho-ing *Resolution* out of the workshop she'd lived in for the past two years and three months. She rolled forward obediently, but we had to take the drop-off from the concrete floor of the shop to the hardpan ground outside into consideration.

Masaya and Stanley laid out five fence posts that John just happened to have lying around so that the 2x12 runners of the cradle would ride up on them as the boat edged outdoors. They worked magnificently. Again, John and I hauled on the pulley while Masaya and Stanley moved the logs as they rolled out from under the sternmost part of the cradle and placed them beneath the bow where the runners would catch them.

Partway out, we changed the direction of the pull and *Resolution* turned her head to starboard as if she were on the ocean. At last, we took the pulley around to the stern and hauled her around a bit more until she'd made a full 90 degree turn from where she left the shop.

All told, the job took less than two hours. And oddly enough, nothing we did invoked Murphy's Law.

When *Resolution* had settled in her place to await finishing touches, John said, "It was like watching some treasured baby animal being born. There was lots of help to get her out into the world, and there are only a few weeks to pass before she is in the water and can move under her own power."

The work ahead of us suddenly seemed less daunting. Soon we'd be installing hardware and fitting the standing rigging. And if I could find enough lead, we'll cast the keel. Lead is a very difficult scrap metal to find in New Zealand. Something I did not know.

Remembering Stanley Fong

I met Stanley Fong in 1988. He'd run up a pile of debts doing business in his homeland of Hong Kong and had come to Japan to gain some respite. Stanley always wore a smile. He was always positive, but he never seemed to have luck with his business efforts.

Within weeks of Stanley's arrival in Chiba, he met the love of his life, convinced her that she was the one, and married her in February 1989. Months later, he took her to Hawaii where, at the age of 34, Stanley began his college education. When he graduated four years later, he had two daughters, with a son and another daughter yet to come.

Stanley and his family became good friends with me and Masaya Kinoshita. It was probably Masaya who talked Stanley into taking that trip to New Zealand to help me and John move *Resolution* outside.

Although Stanley was blessed with his perfect wife, four children he doted on, and many forever friends, he had the misfortune of being diagnosed with pancreatic cancer in April 2011. Intense back pain led to the diagnosis.

As he lay in his bed at the Chiba Cancer Center Research Institute, Stanley told his wife, "I am not afraid. I don't worry about the kids. What I worry about is you." How he loved that girl.

Smiling Stanley Fong left this life at the age of 56. He was always a good friend and though I often visited him in his hospital room, the day we moved *Resolution* outside will always be a special memory. "See ya there," Stanley said. "I'll be waiting for you to show up." Would that *Resolution* could be there, too.

To Cast a Keel

Murphy's Law rules the universe. I'm sure of it. My berth, F21 at Tauranga Bridge Marina, awaited *Resolution*'s arrival. But first the keel ballast must go on. And we hadn't even got it cast yet. Thanks to Murphy.

Let me tell you about it.

When *Resolution* was an upside-down hull with 300kg of kwila hardwood deadwood bolted to the 18mm laminated plywood bottom, I spent days making a blue foam plug the size and shape of the lead ballast. Finished and faired, the foam plug was put aside to await the day it would buried in casting sand and poured full of molten lead. The idea was to pour the lead right onto the foam and the searing heat would reduce the foam to nothing and fill the mold with lead the shape of the foam plug.

Whenever you have an idea, it's good to test it.

Come time to cast the keel, I bought 500 kg of lead (a four-letter word that should be spelled G-O-L-D) to go along with the 300kg or so in tire balance weights that I had collected. I set up a crucible bought on TradeMe (NZ's eBay) over a gas ring burner and used the LPG tank from my caravan to fire it up. I melted down tire weights and poured them into a mold for John's large drafting weights.

Guess what?

The foam didn't disappear. It left a significant residue in the bottom of the trial pour, and we could see that too much would remain in a large project like the ballast keel.

Scratch the idea of using a foam plug.

So. We remembered an article about pouring a ballast keel in a wooden box. I went to Three Brothers on Te Rapa St., and got two 2x6s and 2 2x12s, straight and true as any I could find. The 2x6s measure 142mm, one millimeter wider than the keel on each side. With a 2x6 nailed inside the 2x12s, the box is just 200mm deep. Exactly right.

But there's a lot more to it than just nailing two 2x12s to a 2x6 bottom piece.

It must also be set up for 11 keel bolts. And that means getting a 32mm dowel and cutting off eleven 50mm pieces, drilling 32mm holes in the proper positions along the bottom of the mold, and pounding the 32mm plugs into them. This will make a countersink, as it were, for the keel bolt nuts and washers.

In the 32mm plugs, I drilled 12mm holes in the centers. I put 100mm lengths of 12mm dowel into the holes, then placed 15mm OD copper tubing over the dowels, held by cross bars with holes drilled in them.

I painted the inside with sodium silicate, also known as water glass, and hardened it with a CO_2 fire extinguisher. That would keep the sides and bottom of the mold from burning until the molten lead cooled and solidified.

The mold awaited its fate in the workshop; I dug its trench.

Murphy's Law took effect again.

The idea was to block up an old cast iron bathtub over a fire pit on cut stone, one end slanted down toward the mold, which would be buried in the ground at an angle to the pit. Naturally, the first firepit I dug was in entirely the wrong place. Fill up, start over.

Set the mold. Stand the old bathtub up on blocks of cut stone. Bank earth up against the stone to help the draft run from downhill by the mold to uphill 180cm away.

Murphy struck.

The drain fitting's retaining ring was plastic. Not much use with a fire roaring beneath it. Well, if we stick the drain to the tub with silicone stickum, the whole thing will work, right?

Not on your life.

We fired up the wood and coal beneath the tub, which had 50-60 kg of lead in the bottom. As the tub heated, missiles of porcelain exploded from the inside. I went for a pair of safety glasses.

The lead started to melt. Rivulets rolled down the bottom of the tub, paused for a moment at the silicone goop, slipped beneath it, and dropped into the ashes beneath. I ran for the frying pan I'd used to mold lead from the tire weights. Shoved it beneath the tub to catch that precious gold, er, lead.

Thanks, Murph.

We let the fire burn down.

I got a steel washer from Ross Todd's in Cambridge, reamed it out a bit so it would act as the drain fitting's retaining ring. Friend Ian set up his lathe and turned an aluminum plug to fit the drain. I've put the drain fitting in place and cinched the retaining ring up hard against the tub's drain opening. Tomorrow, Murphy willing, we'll relight the fire.

In the meantime, I worked on the boat outside. I installed the lifelines, the rubbing strips, and the sheer strips. I installed most of the bronze fittings, fore and aft, and then got ready to hang the rudder.

Of Fish 'n' Chips

There's a pub at a yacht club in Olympia, Washington, to which I was drawn while I was working on *DoriKam*, getting her ready for the trip to San Diego and on to Pichilinque in Baja California. But it wasn't the ale called Moose Drool that tempted me to the pub time and again, it was their fish and chips—the best I've ever tasted in the USA.

Months before *Resolution* was trucked to Tauranga, I drove there at least once a month to attend meetings of Tauranga Writers, Inc. My course took me northeast to Morrinsville, where I had to take a right turn onto the road to Matamata (the place where you can catch the bus to Hobbiton). There's a roundabout on the edge of town that lets you turn onto Broadway, Matamata's business street. Driving straight down Broadway puts you on the road to Tauranga, but that's not the point of this tale. It's all about fish 'n' chips.

Who knows why I stopped at a little take-away place called DK's Burger Bar? Who knows why, of all the items on the menu written in chalk on greenboards above the order window, I chose fish and chips? Who knows why, but I did. And I'll never regret it. BK's Burger Bar offers the best fish 'n' chips in New Zealand, I think. What's more, when John and Indy drove me to Tauranga, we stopped at DK's. After 10 years and a change in ownership, the fish 'n' chips are still really good. Totally glad we stopped.

Every month, I and a gaggle of people get together at the Masago Community Center in Chiba for Charlie's Cooking Class. When I went to New Zealand, the members demanded that I bring back New Zealand recipes. I couldn't get one from

DK's, but I did find a New Zealand cookbook with fish 'n' chips recipes.

In Matamata, I chose fish 'n' chips with hoki, a local deepwater fish of the hake family. In Chiba, we used catfish from the USA. The results were better than just good.

Needless to say, I had no fish 'n' chips on *Resolution*. On DoriKam, though, as the sun was rising just a short distance from Cabo San Lucas at the end of the Baja California peninsula, a mahi-mahi hit our trolling line. Breakfast was not fish 'n' chips, but sautéed mahi-mahi is just as good—a fish so nice they named it twice.

More Bits and Pieces

Resolution looked nice with two coats of Energy Yellow paint on her hull. Now the little jobs that had to be done before the launch seemed like a mountain.

One, I put the mighty 5/8" bronze bolts through the three sets of gudgeons and pintles that hold the rudder to the boat. Polishing and grinding with an emery wheel and edge grinder, I got them swinging freely and ready to install. The bolts then went on the drill press, where I bored 3mm holes for cotter pins to hold the slotted nuts in place. One more, small job done.

Then I fitted the pulpit and pushpit. The rails at the front of a boat are called the pulpit because they look like the place where a preacher would stand. Those at the stern were named pushpit for a simple, logical reason. If you pull from the front, surely you push from the back. Hence, pushpit. The stanchion bases got their hardwood pillows. And the tall stainless stanchions waited their turn.

I took all the winches apart to check their viability. The Murrays were ready to install. One pair was questionable. Two other pairs were solid and good equipment. Thanks to eBay and TradeMe. Bronze portholes from the same sources were ready to install. Along with cleats, fairleads, and other small items than help make a boat go around.

The wind vane fit as if custom-made for the boat. Bolt holes, proper alignment, tighten down, and another piece of equipment was installed.

Progress is never as quick as you'd like. But I kept at it day after day, and *Resolution* got closer to completion. Remember, she was a sturdy blue-water boat built largely by one old man in pursuit of the dream of sailing around the world in a boat of his own making.

Another Dreamer

Someone else dreamed of a solo voyage around the world at the time I was building *Resolution*. Her name was Heather Neill and she left Steinhatchee, Florida, in her Flicka 20 named *Flight of Years*, beginning her planned circumnavigation in January 2008. Heather and I became acquainted via the net and she and I conversed a few times over Skype.

Heather is a single mother and former real estate agent who decided it was time to sell her business, buy a Flicka, fit it out for ocean voyaging, and leave for lands unknown. Her route was to take her through the Panama Canal, down to the Marquesas, Fiji and other South Sea islands, through the Torres Straights, across the Indian Ocean, and up the Red Sea to the Suez Canal. She then planned to sail around the Mediterranean, out between the Pillars of Hercules, across the Atlantic, and back home to Florida.

January 1st, 2008, was the day Heather left to sail around the world. On January 3rd, she locked herself out of the boat's cabin and had to break into it through the companionway and hatch boards. She injured herself in the process, which led to her abandoning her dream. She and the *Flight of Years*, were towed back to port two days later. At the end, she wrote on her website, which has since been taken down:

"The weeks since returning have been, and remain, the most difficult of my life. I have not been able – and will not now – relay to you all that has happened. There is no sense in it. It is taking everything I have to write these few words to you.

"My medical bills have escalated and I now face the probability of surgery on my declining hand. My funds are dwindling and I have returned to work in a real estate market which is not currently meeting the bills. I am in the process of relocating to Gainesville, where I will rebuild my life anew, continuing to work in real estate, and probably, for a time, a part-time job as well to make ends meet.

"I am therefore forced to sell *Flight of Years*. I will not itemize here all the costs and work which have gone into her over the last year. Suffice it to say I have countless hours of work and some $100,000 in her (provable by receipt), including thousands of dollars for rigging, equipment, the Monitor self-steering wind vane, the Air-X wind generator, satellite phone, dinghy and motor, as well as $14,000 for custom-built Ultimate Offshore Sails by the German company, Schattauer Sails. You have only to read the months of preparations on this website to see the work, love, and money which have been lavished on her."

Should I have taken Heather's broken dream as a precursor of my own?

"Everybody said, "Follow your heart." I did, it got broken"
--Agatha Christie

Chapter Thirteen

Closer, Ever Closer

"Not far to go now," John Welsford wrote in the *Sundowner Diaries*. "The little boat looks very good sitting out there on the piles of stone that form her temporary cradle. There are lots of bronze fittings bolted to the deck, cleats and fairleads, portholes in the cabin sides (Yes, they came from TradeMe, a wonderful source of bits and pieces). There are chainplates and Samson post pins, winch pads and so on, all bolted down and ready to use.

"From here the boatbuilding gets into much smaller details. Instead of asking for a hand to lift a 16 foot long 7x3inch piece of kwila up onto the keel or the pair of us grunting on the end of a four-part block and tackle to pull a three-quarter ton lead casting into place, it's more about using a dot of sealant to hold the washers on the back of the bolts until a nut can be done up, or visualizing the fall of a halyard to get the cleat in the right place. As with every stage of the job, the list of things to do seems to grow faster than they get ticked off, but the jobs are getting smaller all the time, and fewer.

"Right now, the task is to prioritize the jobs into what should be done "here" while the workshop is available, and those that can be done on the hard or in the water at the marina. Yes it's that close. The truck is booked, and we expect to see that bright yellow transom head out the gate soon. For two and a half years it's been a part of our lives, gradually changing from a pile of lumber that could just as easily have been firewood into an ocean-capable voyaging home, the culmination of this part of a dream."

Charles T. Whipple

How Slowly Doth Progress Occur

Already it's been two months since we cast *Resolution*'s lead keel in a wooden mold of my own making. Just yesterday, I spent a day putting in four keel bolts.

Why does it take so long?

Nothing seems to come together in seamless progress. There's always a missing part or a drawing not finished or a tool not available.

Take the ballast keel, for example.

When it came from the mold, the keel was a squarish lump. Imagine a scow about 12 feet long, six inches wide, and eight inches deep. That's the ballast keel. The problem is, the deadwood is a different shape, especially in the forward section where the wood is rounded to direct the flow of water back along the keel.

I measured the deadwood notch where the ballast was supposed to bolt on, and then measured the ballast. Too long by a long shot. Something like four inches longer. John gave me an old handsaw, and I cut off the extra lead. The scow now had a transom bow.

As I mentioned, the deadwood is rounded at the forward end. So I made a template of the shape and traced it onto the lead. Out came the old saw, and I cut away the extra lead. But now that I get the lead ballast on the keel, I see there will be more shaping of lead to do. Under that heavy boat, too.

Anyway, the time came to flatten up the upper surface. I got out the 10-foot length of aluminum we use as a straight edge. Immediately, I could see that there was nothing straight about that hunk of lead. It was sway-backed as an old mare, sagging nearly an inch between ends. Not only that, but it curved markedly to port, stern to stem. Egads. How do you straighten out three-quarters of a ton of lead?

One-pound Hammer

John tells a joke about a young university researcher whose field of expertise was medieval history and lifestyles. She heard of a village in Ireland that maintained its traditions over the centuries, and she determined to visit the village to sample its aura of bygone years.

The researcher arrived at the village near dusk after flying across the Atlantic, taking a bus to the hinterland, and then a horse-drawn cart to the village itself.

Indeed, the vista was straight from the Middle Ages. Daub and wattle cottages with thatched roofs. Fields lined with walls of rock. A solid church at the town square. And a village smithy.

The woman had written the smith, who was famous for his strength and skill, and had arranged to board with his family during her stay in the village. She thanked the old man who had driven the cart, took her suitcase in hand, and made for the blacksmith's place of work.

Inside, the forge glowed as the smith's helper worked the bellows. The smith, a giant of a man, held the work to the flames with a huge pair of tongs. When the iron reached a cherry red, he pulled it from the forge, laid it on his anvil, picked up a hammer from a box of hammers lying at hand, and delivered the iron a mighty blow. He then tossed the hammer out the window onto a growing pile of hammers, and thrust the work back into the flames. When it was again cherry red, the smith selected another hammer, delivered a single blow, and tossed the hammer out the window.

That night, after a hearty meal and innocuous conversation before the hearth, the researcher finally asked the question that had been bothering her since she'd seen the smith at work.

"Tell me, Paddy," she said. "Why do you use your hammers to strike only one blow?"

"Faith, and I thought you'd notice," the smith answered. "They're all one pound hammers."

12-pound Sledge

Examining the curved hunk of lead with his "eye-crometer," John said he thought we could beat it into submission with a sledgehammer.

No sooner said than done. I jumped in my trusty Mitsubishi and rushed off to Placemakers. The biggest sledge they had was a 12 pounder. I bought it.

Back in the workshop, I used a long pipe and a length of chain to turn the lump of lead upside down. Eight thwacks with the sledge, and the sway was out of the ballast.

I turned it on its side, bowed portion up, and another four thumps to get that part straightened out.

Notice that we used all twelve pounds in the sledge, so we set it aside to recharge.

Now all I had to do was smooth off the lumps.

You see, we poured the lead so that the top of the ballast keel was open, so the surface needed to be smoothed. John suggested using the power plane.

The power plane in the shop had seem many years of service and shaved uneven surfaces of thousands of board feet of lumber, ranging from softwood Fijian kauri to Douglas fir to hardwoods such as jarrah and kwila. It lasted about half an hour chopping at the lead before the main bearing burned out.

I bought another, a local hardware store's generic brand, for NZ$40. It, too, lasted about 30 minutes.

Forty dollars for another planer. But this one lasted the whole job.

Still, planers are designed for flat surfaces and the front of the ballast must be rounded.

"The thing to move lead with in a hurry," said John, "is what they call a panel beater's rasp. Don't have one."

In the States, we'd call them sheet metal worker's rasps, but be that as it may, it just so happened that our local garage had a set they don't use every day. Mike, the owner, has been watching *Resolution* take shape, and offered me the use of the rasps. I accepted.

A couple more hours and the bow end rounded off nicely.

Now all we had to do was wait for a break in the weather and in John's schedule to roll the ballast outside and slip it under the boat.

Of course, I kept busy while the lead waited. I put the mainmast together. Sanded and painted the boom and the gaff. Laid out the anchor chain winch and pipe hawse hole, set up the bowsprit, fitted the bobstay and whisker stay fittings, and placed the stays. Put access hatches in the stern seat top so I could install the through-transom exhaust fitting. Hung the rudder. Assembled and painted the boom crutch. Took 180-plus feet of chain off to be re-galvanized. So you see, it seems the ballast keel takes forever to get in place, but that doesn't mean nothing is happening.

Oh, I installed all the rubbing strakes and bulwark strips on the boat as well.

We rolled the lead outside on fence posts, and noticed a peculiar thing. The lead had a sag in it again. The only thing we could figure out was that the lead sagged under its own weight while supported by two balks, one at each end.

Again, weather interfered, but the second shiny day after that, we got out the big come-along pulley, parked John's car in the line of pull, hitched the come-along to the trailer knob, and pulled the ballast keel up under the bow of the boat. A couple of pipe levers straightened the lead (not the sag, the positioning) and we used the come-along to pull it backwards into position.

It took me the better part of a day with the hydraulic jack to inch the ballast into precise position, but when it got there, it fit almost perfectly (except for the sag off to the starboard. We'd fix that later.).

Splicing the Wire

Resolution was designed to be a simple boat. Virtually anything and everything could be repaired while underway . . . at least that was the concept.

Part of that simplicity was the rigging.

When I started building *Resolution*, the plan was to make the spars, including the mast, of old-growth Oregon pine. But when the time came to build the mast, we could not find the right wood, not any that fit the building budget, anyway. So John drew plans for a mast made of an aluminum flagpole blank, 7.2 meters long.

On top of the mast, a T-shaped hardwood masthead crane brace and light platform that served as a seat for the tricolor light that would tell the world that *Resolution* lived and sailed the ocean brine. In its forward flange, it had holes for the flag halyard and the light weather topsail halyard blocks. And in the after flange was a hole for the drifter halyard block. The crane itself was made of hardwood 20mm thick and fastened to the aluminum pole with 6mm stainless bolts. I attached hounds for the forestay and the forward and after side stays at a point 110cm down the pole. Another 85cm further down the mast came the staysail hounds, which supported port and starboard staysail side stays, a block for the gaff throat halyard, the staysail halyard block, and the staysail forestay.

From top to bottom, *Resolution*'s mast measured 8.5 meters.

Resolution sat outside the workshop, hatches closed, and her complement of fitted pieces of bronze grew day by day.

When it rained, which happens with regularity in New Zealand, I worked on rigging, built the mast, and finished up the gaff and boom. All quite straightforward now that I'd had nearly two and a half years of experience in building boat things.

Ah, the mast tabernacle. Made of 4-5mm mild steel plates, then thoroughly galvanized. It fit between the lockers in the forward end of the cabin and was bolted securely to the compression posts that also made up part of the galley's right-hand wall.

Actually, building the spars turned out to be the easy part. I was used to working with wood and epoxy and completely agreed with the third Welsford theorem of boatbuilding. "There has yet to be a boatbuilding error that cannot be remedied with epoxy and bits of wood." Unfortunately, the third theorem applies only to building boats of wood and epoxy, and does not apply to splicing galvanized wire rope into rigging stays.

The forestay and jib stay were pieces of cake. I measured out 6mm stainless steel wire, cut it to length, and swaged a loop in the top and a thimble in the bottom. Simple stuff done many times before.

The side stays were a completely different thing. They were made of 4.5mm 7x7 galvanized wire rope, but setting them up was much different work than just measuring and swaging.

All the stays had a large loop in the top end and a small one with a thimble in it on the bottom end. Both had to be spliced into the wire rope. Spliced. Not cut and swaged. Spliced. A whole new thing for me.

So how did I learn to splice galvanized wire rope?

Just like I learned everything else while building *Resolution*—trial and error, with a little preliminary study to tilt the balance away from the error side.

First the tools. A fid—which looks like a spike with a handle. It's used to create a gap between strands. A hacksaw to cut the wire rope, or one of its strands, as necessary. Twine or wire to whip the finished splice. Tape to put on the ends of the strands to keep them from unraveling as you splice.

Simple, right? Not in the least.

The descriptions for splicing make it sound deceptively simple, too.

First they tell you to determine the size of the eye you're going to splice. Well. That makes sense.

Second, you apply whipping about three and a half turns of the strands away from the end of the wire rope. If I remember right, I used plastic tape for this and cut it off after the splice was made. You also tape the ends of the strands securely so you can thread them through the openings you make with the fid.

Unwind the strands. The whipping will stop the unraveling after three and half to four turns.

Take a deep breath.

Now. String your wire rope stay from a high point above the vise you're going to use to hold the loop tightly in position.

Let's pretend you're going to be splicing an eye for a thimble in your stay, as I did for every stay on *Resolution*. You wrap the stay around the thimble with the unraveled portion coming just at the thimble's throat. Now put the thimble and wire rope in the vise, positioning it so the unraveled portion is just above the vise jaws. The long portion of the wire rope is held up because you strung it from a high point. The wire rope should now run straight upward and be securely fastened, because you're going to be using your fid to pry strands apart and thread the unraveled strands through them to form a firm splice.

Separate the unraveled strands so there are three on each side of the standing wire rope. The core strand will stick out until your splice is almost done. Take the nearest strand on the right-hand side, use your fid to open the standing wire rope as close to the vise as possible. Push the closest strand on the right-hand side through the fid opening in the center of the standing wire rope.

Now take the second right-hand strand and push it under two strands of the standing wire rope. The third right-hand strand gets pushed under the remaining strand on the right-hand side of the standing part.

You have three strands on the left side of the standing wire rope, too. Treat them exactly the same as the right-hand ones. First strand (which is actually the fourth strand) gets pushed through under two strands. The fifth and sixth strands are treated like the second and third strands were, but in mirror image.

Pull all the strands tight and hammer them down, using either a hammer or a spike.

All you have to do now is continue the same process, moving in a spiral around the standing part of your wire rope, five times in all.

At the end of the fifth round, you cut off every other unraveled strand and whip them. The remaining strands are spliced into the standing part one more time, cut off, and whipped. The core wire is also cut off with enough protruding to allow it to be tucked into the standing part.

You'll probably want to worm the splicing before parceling and whipping it. In my case, we boiled the stays in a linseed oil and Stockholm tar mixture and hung them out to dry, so to speak. Then I wormed the splices, parceled them, and slushed them with the oil/tar mixture again. The bottom splices held thimbles and needed no other reinforcement, but the loops at the top, which fit over wooden hounds, were covered with leather for additional protection.

After they'd dried and the tar/oil slush had hardened up, I coiled the stays and put them away against the day when *Resolution* would be in the water and I'd be raising her mast and rigging.

On the Move

Once the ballast keel was bolted on and rasped into some semblance of shape, the boat was basically ready to ride a specialized low-boy trailer to Tauranga for her launch. I asked the marina to recommend a hauler, and Julie, who worked under Fred Jeanes, said Peter Jacob was the man to talk to. I called Peter and got a quote from him. Lots of money, but what could I say? John and I discussed all that had to be done, and decided that April 4th would be the right day for *Resolution* to change addresses. I emailed Peter a week in advance, asking him to do the honors.

No answer.

Finally, on the Monday before Friday April 4th, I called Peter. "No way I can haul your boat on Friday," he said. "And I can't do it next week either. I'll let you know."

On Monday the 14th of April, Peter called and said he could make the haul on Friday the 18th. We, of course, agreed. Getting a truck set up to haul yachts is no simple matter.

At 11 a.m. that Friday, a huge crane trundled into John's graveled driveway, across his lawn, and around back to where it could reach *Resolution,* which sat on piles of cut stone. The driver used balks of timber to make footings for the legs that stuck out from his truck to steady the crane. Then we looped the strops around *Resolution* as Peter backed his truck over John's lawn.

I tied a polypropylene rope to the pushpit rail and held *Resolution*'s head to the wind as she flew from her bed of stone over a pile of fencing to slide demurely onto the trailer of Peter's yacht hauler. Within minutes, *Resolution* was rolling out of the driveway and down the access road toward Tauranga. She looked small, but very brave, standing proud and all alone on that big trailer.

But my ocean-going craft first had to traverse the miles from Hamilton to Tauranga before she could be given several coats for anti-fouling paint on her shapely bottom and set into the waters of Tauranga Harbor. And there were miles to traverse and work to be done before she could sail away on the first leg of her trip around the world.

There is always a sadness about packing. I guess you wonder if where you're going is as good as where you've been.
--Richard Proenneke

Chapter Fourteen

Do You Know the Way?

The way to San Jose? Not hardly. The way to Tauranga.

You've heard the name. It's where *Resolution* first wet her keel. But long before that happened, I made my way to Tauranga on the first Thursday of every month. You see, I was and am a member of Tauranga Writers, Inc., the oldest and perhaps most active self-help writers' group in all of New Zealand.

So once a month, for more than a year before *Resolution* was launched, I drove my venerable Mirage to Tauranga to attend an evening TW Inc. meeting.

The route took me through a town called Matamata, where a little takeaway place known as BK Burger Bar offered the best fish 'n' chips in the whole country. In my opinion, of course.

It took me a couple of hours to drive from Hamilton in the middle of New Zealand's North Island to Tauranga, which lies on the Bay of Plenty and has a good deep-water harbor. Later, when I moved aboard *Resolution* at Harbor Bridge Marina, I learned the way to McDonald's and Burger King up the hill and to Starbuck's over on The Strand. Favorite places for breakfast (I couldn't drive to Fran's Cafe and I found no alternative to her place in Tauranga) and a tall cup of Signature chocolate of an evening. Still, dawn till dark, I was at the marina working against the clock to get my boat in condition to sail the first leg of my circumnavigation, which I had now decided was to Hawaii.

I get ahead of myself.

Once TW Inc. President Jenny Argante asked me to conduct a focus session on *haibun*, a Japanese literary form that encompasses not only poetry but also just about every form of prose. I, having taken a class in the art, said, "Yes." I prepared and gave the workshop, only to find out that one of the members, Kirsten Cliff, was a haiku poet of some renown. In her mind, she must have been rolling on the floor with laughter at my clumsy presentation. Outwardly, she acted like any other session attendee. True Kiwi friendship.

But the TW Inc. member who's still a best friend forever was Janice Giles. She christened *Resolution*. She's stayed aboard, even when I've wanted to abandon ship. Always there, always a friend, even half a world away.

When I made my pilgrimage to New Zealand in March 2018, one person I vowed to see was Janice. She's somewhat of an Amazon, having crewed on a commercial fishing boat, among other Herculean efforts.

John, Indy, and I drove up to Janice's house without even minimal fanfare. Not only is Janice a bit of an Amazon, she is also the creator of stunning paintings. After a guided tour through her new place, we settled down to talk. But first, Janice went over to a cabinet and came back with the cork from a bottle of sparkling cider. "Here," she said, holding it out to me. "It's from *Resolution*'s christening."

I have it now, along with the porthole from Scott and Isabel Mabey and a piece of Resolution that John rescued from the wreck site.

In a way, Janice reminds me of Patrick Wong, another forever friend.

I worked for H.J. Heinz Company then, first as assistant to the President of the Japanese joint-venture company, and then as North Asia Export Manager, based in Hong Kong when it was still a British colony. My office was in a corner of the Muller & Phipps premises. M&P was the Heinz agent in Hong Kong, and Patrick Wong was a sales manager there.

Sundays on the golf links were practically mandatory in Japanese business, but businessmen in the Colony flocked to the Saturday horse races at Happy Valley on Hong Kong Island.

Our little clique finally got honed down to myself, Patrick, and Alan Chan, a man who ran a small printing company. On Friday nights, we gathered at Patrick's house, armed with the little magazines that charted the performance of every horse in every race for the past year. On those Friday nights, we made our betting decisions for the next day.

My own Saturday method was to take HK$100 in $10 bills. Those bills were located in my right-hand pants pocket. I would bet one bill on each race. If my horse won, I put the take in my left-hand pants pocket. I never used any of these winnings to buy tickets on later races.

One Friday, we discussed the horses in the final race much more than usual. None seemed to stand out. In fact, all seemed to be culls, horses that had not won or placed the entire season.

Alan heaved a sigh. Anna, Patrick's wife, stuck her head in the little study room where we devised our betting strategies and said, "Coffee anyone?"

All three of us raised a hand. Anna disappeared back into the kitchen.

"Hey," said Alan. "Look here. This horse, this filly's name's Anna. My money's on her."

Patrick and I exchanged glances and nods. "Sounds good," I said. "No other nag stands out. I'll go with Anna."

"Okay. Me, too," Patrick said.

Saturday did not see any of us win a lot of money. Down to the final race. As I remember, Anna was a bay filly with black points. Nothing set her off from the rest of the field, but we'd agreed to bet on Anna, and bet we did.

With us watching then leaning forward as the field rounded into the home stretch, there was Anna, tail streaming and neck stretched toward the finish line. Anna. Anna, who had never won a race in her entire career, moved ahead by a neck, then half a length, and by the time she plunged across the finish line, she was nearly a full length ahead of the runner-up. She won! And the odds were something like 13-to-1. For the first time in months, my left-hand pants pocket bulged.

Maybe Murphy's Law slacks off sometimes at the Happy Valley horse races.

Patrick moved to Canada when Hong Kong reverted to China. I don't know what happened to Alan, but Patrick probably does.

On the Hard

I arrived at Tauranga Bridge Marina about 2:30 p.m. Peter had the boat unloaded, sitting on her ballast with wooden blocks beneath her belly. Well, strictly speaking, Bruce unloaded the boat with the Travelift, but Peter waited for me to arrive as if he had done the entire job alone.

"Where ya been?" he wanted to know.

"You said I didn't need to be here when you arrived," I countered.

"I didn't mean waiting for an hour and a half, though," he groused.

I apologized. Took the invoice, and promised to have the money to him yesterday, or as near to yesterday as I could possibly manage. He drove away in such a huff that he left his bow strops hanging on my boat.

A deep breath helped clear the air. Then I walked down to the office to talk to Bruce.

"You're Charlie, eh? OK. We'll move your boat into the place that the little launch will open up tomorrow. Then you can get to work on her."

Still, *Resolution* was not in her designated position on the hardtop until just after noon on April 19th. After lunch that day, I started work.

The paper masking tape that marked the waterline had been adhering to *Resolution*'s hull since late November 2007. Masking tape is designed to pull off the work easily, leaving a clean line of paint behind. But you're not supposed to leave the tape stuck in place for months.

I pulled a small hand scraper from my tool bag, and started stripping the masking tape from beneath *Resolution*'s Energy Yellow topsides. It took me three hours plus to get it all off. By the time I had laid another line of masking tape 50mm above the true waterline, the sun was threatening to disappear in the west. Besides, I didn't have a sanding block.

By 8 a.m. the next morning, I did have a sanding block, sheets of 100 grit sandpaper, and lots of elbow grease. By noon, the bottom was sanded, wiped down, and ready to paint with a coat of International epoxy primer. Now, the first coat of antifouling paint was supposed to go on within six hours of the primer coat. So that day, April 20th, I finished the first coat of Shark White No.10 hard antifouling after the sun was already down.

April 21st saw me with roller in hand, putting a coat of red No.5 ablative antifouling paint on *Resolution*'s bottom. I followed that coat of paint with another, six hours later. In two days, I covered her bottom with four coats of paint. I patted myself on the back, then drove back to Hamilton for another carload of necessities, parts, and other things that *Resolution* and I would need. After three nights sleeping in the boat, a night in the caravan was pure rest. Bridge Marina lies alongside the main artery for trucks heading from Tauranga commercial harbor to the rest of the North Island. Loud . . . well, anyway.

Back at the marina with my load by 10 a.m. I stripped the masking tape from the waterline. Strange how well it comes off when it's only been in place for a couple of days. Laid another line of tape around the upper edge of the waterline, then another above that, leaving 100mm at the forward and after ends and narrowing to 60mm along the center of the boat. By the time John arrived at 2 p.m., the boot-top had been painted with two coats of black antifouling.

We'd planned to put *Resolution* in the water that afternoon, but Bruce the lift operator suggested we wait for slack tide, which occurred at 9:00 the next morning. Thanks to that wait, Janice was available to do the christening honors on April 23rd, 2008.

"I Name This Boat Resolution"

While we were some weeks after our original target date for the launch, *Resolution* at last kissed the water with the stern edge of her ballast keel.

Moments before, Bruce Goodchap, the operator of the Travelift at Tauranga Bridge Marina, lowered the boat to the point where her bowsprit stood at shoulder height to the lovely woman who had been asked to christen her. The group from WOS, dressed in normal togs rather than club uniform, held its collective breath, each with a crystal plastic cup in hand.

"I name this boat *Resolution*," Janice Giles said as she poured bubbly apple cider (*Resolution* is a dry boat) over the cranse iron. "May God bless her and all who sail in her."

Ship of Dreams

John poured cider for all assembled, and we toasted the boat and her maker and master. Bruce moved her out into the dock, lowered her to deck level, and I clambered on, followed by John. The boat settled primly into the water. I stepped below and turned the starter key. The little Bukh engine happily rumbled to life and shoved the boat forward and back as John worked the morse lever.

We backed out of the Travelift's strops and into the wind blustering from the south. The Bukh sputtered, coughed, sputtered, coughed, and died. She wasn't getting fuel. Andrew came roaring up in an orange inflatable with a 115HP outboard doing the pushing. In moments he had us in tow and headed around the breakwater and into the south section of the marina toward berth F21. He idled the inflatable and allowed the wind to blow us north and east, stern first into the berth where *Resolution* made her home until we weighed anchor for the Sandwich Islands.

Windsheer

Fred Jeanes was the man in charge at Tauranga Bridge Marina. Not only did he run marina operations, but he also owned a slip where he moored the lovely *Windsheer*. She must have been 36 feet LOA and she was pristine. I never saw a cleaner yacht. Fred let me board her one day, and a delightful experience that was. Sipping on a cuppa, we talked marinas and boats and construction of *Resolution*. It was almost like *Windsheer* was a sister ship to *Resolution* herself.

Of course, Fred was a must-see friend while on my 2018 pilgrimage to New Zealand. Fred's is a hillside house. The day was a sunny one, almost as sunny as Fred's perpetual disposition.

"C'm in, c'm in." Fred's voice is a big one and full of good cheer. He led us out to the veranda overlooking a small valley and set us at the table with glasses of iced tea or water, whichever John and I wanted.

He sat with his usual belly laugh and pulled a white folder from under a pile of stuff. He shoved it at me. "Thought you might want these," he said.

I took the folder and opened it. Inside, picture after picture of my *Resolution* as she sat tied up at berth number F21. "Took those when I had a spare minute," Fred said.

It was hard to express my thanks. As usual, the unreserved friendship of the Kiwis I knew made it even more difficult for me to respond.

"Remember well," Fred said, "going down to the lunch room for a morning coffee and there'd be Charlie, writing something, never knew what, but something. Busy writer, you." Fred's belly laugh sounded again.

True, Fred's retired from managing Tauranga Bridge Marina, but he's still hale and hearty, greeting stories in the conversation with, "Gwan now. Gwan."

He also developed a line of heavy New Zealand mariner's sweaters, thick woven with a tail piece that hangs down behind, making sure the wearer's got a warm place to sit, always.

I bought one.

We drove away after a couple of hours of camaraderie, and I remembered how Fred's voice had come back to my call over the wireless. "*Resolution, Resolution.* This is Bridge Marina. Been good having you..."

Now What?

The first nail was driven in the jig upon which *Resolution* was built the first week of September 2005. She left the premises where she was built on April 18th, 2008. Although not yet rigged or fully finished inside, she's afloat, two years, seven and a half months after I started "building" her.

She took a lot more time and a lot more money than I originally thought she would. Perhaps if I had known at the outset the cost in time, money, and missing digits, I would not have started. And to think. Nearly 1,000 days in the making. Yet it would have taken only about 300 sailing days to complete her circumnavigation. No wonder most circumnavigators buy their vessels or have them made professionally.

That night, *Resolution* sat quietly at her berth, moored to four quarters and hardly moving. Aboard, I felt little movement. There was not enough ripple in the water of the yacht harbor to make any sound against her hull. Frankly, it didn't feel much different to sleep on her now than when she was on the hardtop, so steady was she.

In a week she would be fully rigged, with her sails bent on. Which day we'd go out for a shakedown sail, I didn't know, but it would be soon. A little later in the month, I planned to take her offshore. At that time, I was thinking about going to Great Barrier Island . . . if only I had known.

Charles T. Whipple

Take a Haiku

Activities with TW Inc. included walking the Haiku Pathway in the town of Katikati, which boasts the largest collection of haiku-inscribed stones outside Japan. We converged on Katikati in a bevy of automobiles, including my Mitsubishi Mirage. We started off along the path more or less together, but were soon strewn apart like wind-blown leaves as some stopped to examine a haiku carved in stone while others continued on their way.

*stationary bus
talking we visit places
within each other*

--Junice M. Bostok

The pathway is not linear. It wanders down the west bank of Uretara Stream, crosses then wanders back up the east bank. More than 30 monoliths now bear English haiku engraved into their stony skins. An hour or two or three spent on Katikati Haiku Pathway transports one out of the humdrum of daily life and into the introspective world of those observed through haiku poets' pens. Some say it is a voyage of discovery, because each poem is interpreted by the heart of the person reading it.

According to Katikati poet Catherine Mair, "Haiku are words which sing, words which paint pictures, small stories that expand each location (in the Path), images that invite you to make up your own stories, poems which are the direct experience of the moment, tiny poems which are wonderfully large."

I wondered if my tiny boat would paint a wonderfully large picture of the place we call home—Planet Earth.

How inappropriate to call this planet Earth when it is quite clearly Ocean.
--Arthur C. Clarke

Chapter Fifteen

Behind Tauranga Bridge

Resolution sat in berth F21 as if she owned it. Her Energy Yellow topsides made her stand out, a ray of sunshine amongst hulls of dingy white, navy blue, and the occasional dark red.

Below, she was a riot of items that needed stowing, hooks and netting that needed installation, stacks of books for the library in Stanley on the Falkland Islands, tools waiting to do their thing . . . . more than two and a half years since I spent two rainy weeks laminating beams for *Resolution*'s decks. As she sat comfortably in slip F21, you'd have a hard time imagining all the little parts that came together to create such a bonny, buxom lass. The sight of her was enough to make a man's chest expand with pride and make him champ at the bit, eager to get underway.

Rates for visitor slips at Tauranga Bridge Marina are now NZ$450 a month, but I remember *Resolution* being berthed for under $400. After all, it was more than a decade ago.

The harbor has 500 slips, and you can buy one if you're planning on living aboard your boat, or ashore somewhere near Tauranga.

Tauranga Bridge Marina's a fine place to moor a boat. No hoity-toity yacht club, just a staff of friendly people who do their level best to provide for the needs of the boat people.

The marina's main building housed a small diner where you could either eat in or take out, as the occasion demanded. I often had a simple breakfast there, and stretched out the meal writing articles and chapters in western novels, many of which are still among the novels in the WOS collection.

Rigmarole

I'd made the mast of an aluminum flagpole blank. Atop the pole I put a hardwood crane, then hounds for the stays in two places down from the peak. A big piece of laminated hardwood accepted the bottom end of the pole and fit into the mast tabernacle, which was bolted through the deck into the mast compression posts. I built that mast at John's workshop in Matangi and it rode proudly atop *Resolution*, the bottom end lashed to the pulpit and the peak sticking out behind, resting on a boom crutch I'd made without much forethought. Remember Welsford's Third Theorem of Boatbuilding? The mistake that cannot be fixed with wood and epoxy has yet to be made. It is accompanied by the axiom that no mistake will ever be found before the epoxy hardens.

We'll get back to the theorem later.

We'd fit the stays to their hounds while still in Matangi (yes, John helped once in a while). I'd carefully measured and spliced and wormed and parceled and sealed all the stays with Stockholm tar and linseed oil. They say stays built like that last virtually a lifetime.

With *Resolution*'s stays, I didn't get a chance to find out. They say a boat is like a mistress, always patiently waiting for her man to come around. *Resolution* did everything I instructed her to, and ended up on the rocks from which I was not skilled enough to rescue her.

Back to setting up the rigging. Everything about *Resolution*'s rig was oversized. True, she was only 21 feet LOD (not counting a three-foot bowsprit) and might fool the casual observer as to the forces that would come to bear while sailing for a month or more between ports of call. You've got to remember that *Resolution* weighed nearly three-and-a-half tons when we motored past Mt. Maunganui and out the mouth of Tauranga Harbor, bound for the Hawaiian Islands.

One man, even an old duffer like me, could stand the mast up in its tabernacle. All the stays were fitted over their proper hounds and all that needed to be done was attach the lower ends of the stays to their turnbuckles and tighten them down.

Remember Murphy's Law? If anything can go wrong, it will. It did.

Every one of the side stays that I measured and cut and spliced fell short. And not just a couple of centimeters short, either. A couple were at least a foot short, two were more than a foot short, and two were short by a whole two feet.

What to do?

Think.

Sometimes thinking is tough for a geezer who's already past the mandatory retirement age of 65. But I came up with a simple, if not elegant, remedy.

Down to the Burnsco I went in my dull beige Mitsubishi Mirage, in through the double doors (after I parked the car) and over to the reels of chain. I picked 1/4" hot-dip galvanized chain and had the staff cut six lengths to the measurements I'd painstakingly taken of the distances from stay thimbles to turnbuckle eyes. Hot dipped galvanized chain is not as expensive as you might think, and this time the lengths were correct.

While I worked at getting *Resolution*'s rigging squared away and ready for deep water sailing, The boat engine mechanic Fred recommended got down into the bowels of my ship to remedy whatever it was that had stopped the little Bukh engine less than 50 meters from where she entered the water.

I remedied the short stay problem with my bits of galvanized chain, and the mechanic remedied the Bukh problem, but only after draining the Bukh's fuel system to remove the water. Yeah. Water.

The fuel tank held 65 liters and when the motor got its test run, only a liter or two of fuel went into the tank. But before launching *Resolution*, I took a yellow 18-liter polytank to get diesel fuel. Thing was, the polytank had been sitting in the cockpit with the lid off, and I didn't think to check for water. As it didn't weigh more than it should, so the amount of water was miniscule.

The little Bukh was just what it sounds like, a small diesel weighing only 62kg, so a little water stops her up completely. I worked the hand lever on the fuel pump and the glass separator bulb showed bubbles in the incoming fuel. I thought it was air, but no, it was water.

Once water gets in the tank, it settles to the bottom and goes right to the lowest point. That's exactly where the Bukh's fuel pump attached to the fuel tank. A little water goes a long way, and it took some time to drain it out, along with several liters of diesel.

Still, there was nothing wrong with the motor and once the water was drained out, the little Bukh started right up and chugged contentedly like the happy youngster it was.

Measure, Re-measure, Then Measure Again

The maxim in boatbuilding by amateurs is measure, re-measure, mark and measure again, then make the cut. Still, according to Welsford's Third Theorem of Boatbuilding, the mistake that cannot be remedied with wood and epoxy has yet to be discovered. Ha.

When setting the stays on their hounds, I also fit blocks for jib/genoa, staysail, gaff throat and peak, and courtesy flag, and threaded the proper halyards through them. The mast looked like a second cousin to a maypole when I stood it up in its tabernacle. Still, after all the stays were tightened up and the halyards coiled and stowed on their proper hooks, *Resolution* looked much closer to shipshape and Bristol fashion.

Then Murphy's Law struck.

Resolution's boom was a long stick that stuck out past the transom even when it had the mast firmly in its jaws. No problem. The boom crutch, a red and yellow piece of woodwork of my own design, held the boom right in place. All I had to do was lace the sail's foot to the boom, the luff to the mast, and the head to the gaff. To make sure everything fit right, I hauled the gaff up the mast with its two halyards. It went up like a charm, all the way to the lower hounds.

That should be it, right?

Should have been. But wasn't.

The boom still sat on the crutch and at least a couple of inches of sail ballooned at the foot. Murphey's Law indeed.

Take care of any mistake with wood and epoxy, says Welford's Third Theorem of Boatbuilding. So I set about lowering the mast jaw seat by six centimeters. Once that job was completed, the sail's luff was taut from boom to gaff, just as it should be. But the boom crutch was still too high. The main's boom would not swing. Remedy? Get rid of the boom crutch.

Done.

But what to do with the boom and gaff and furled main sail? Without a boom crutch, they'd drop right into the cockpit and be a huge bother while *Resolution* was in port. Of course the topping lift could take them out of the cockpit, but the boom would still swing back and forth.

What to do?

Think. Wonder. Think again.

Then I noticed the flanges on the pushpit poles. The boom extended out past the pushpit and was in fact resting on the port side pushpit rail as I puzzled over what to do.

The flanges were a good 5-6cm wide and looked to be 2.5cm thick, put there no doubt to hold fittings for the stern gate, if ever I decided to mount one.

Instead, I used them to support pieces of the old crutch, one on each side, as legs for a new crutch. And wonder of wonders, the boom rested naturally on its new crutch, up high enough to be out of the way, low enough to make furling the sail and lashing the gaff and sail to the boom relatively easy work.

Testing, Testing, Testing

The day came. A banner day. The day when *Resolution* backed out of her slip, forward between the rows of berths, through the yacht harbor entrance, and out into Tauranga Harbor. For the first time in her life, *Resolution* was out on her own.

Outside the narrows but still inside the harbor, we powered down the little Bukh and started hauling up the sails. John accepted the job of hauling the sails aloft as if he'd done it a thousand times before. Ya think?

From John Welsford's *Sundowner Diaries*. (The entire clutch of them can be read on John's jwboatdesigns.co.nz website):

"I was concerned that turning to starboard with a long keel and a clockwise prop rotation might make her reluctant to come around but that big rudder brooks no argument and she turned easily within the confines of the channel between the slip fingers. For a long-keeled boat she is quite nimble, like a short and chubby lady square dancing and showing that she is much lighter on her feet than everyone expected.

"Throttle up, and she gently surged away, winding out of the marina and against the tide as we headed down harbor toward

Pilots bay just inside the harbor heads—the 7.5HP Bukh ran smoothly and seemed to have plenty in reserve, and the prop bought on TradeMe did its job really well.

"I sat up on the foredeck while the proud new captain drove, standing at the tiller, easing the boat across and along the edges of the shoals to keep out of the commercial area of the channel and to get out of the tidal current. Nice, it was a perfectly clear winter day with almost dead flat calm and only the tiniest breeze, there were a very few sailing boats further out in the harbor, but they were not going anywhere much and I suspect that those who were had the iron topsail on.

"As we motored off past the containerships and freighters, several large pleasure boats went past, making that horrid square and lumpy wake that planing hulls do at displacement speed and threatening to toss us about. But *Resolution*, designed to be self-damping in both roll and pitch, nodded once to each wave then steadied and carried on, promising that other and bigger waves would be unlikely to make her uncomfortable.

"As we left the wharf area for the more open part of the lower harbor, walking about the decks I unrolled the jib and cleated the sheets off, and a few minutes later hoisted the staysail and sheeted that off. She heeled a little, and seemed to want to move on, the engine note lightening a little as the load came off.

"I took the peak and throat halyards, and hoisted the main, cleated them off and eased the main slightly, then signaled to Charlie to switch off. As the putt-putt of the little diesel faded away, *Resolution* seemed to sigh, come alive and step out a little. Magical. The smile on the skipper's face was just great!

"While there was very, very little wind, she seemed to answer the helm well, tacking reliably and with surprisingly little loss of way, the sheets and controls coming easily to hand, so I abandoned ship with the camera, climbing over the stern and dropping into the little white dinghy with its 2HP Suzuki outboard (and oars, insurance!) and, after starting the motor, casting myself loose. I figured that it would be no problem to catch up if being single-handed for the first time were an issue, but catching him was not as easy as I thought, and sailing the little ship alone was as comfortable and easy as it could be.

"I sat and shot photos as Charlie sailed away, the boat heeling slightly and moving off, leaving just a faint glossy trail rather than a wake as such. I sat waiting with the camera as he came back after going about at the end of the beach. I shot pics, more pics, and more still as she approached, and decided to motor after her and around the other side, but found that it was a real effort to catch her with the Suzuki roaring away and the little white inflatable making a really foamy wake. *Resolution* was as good at sailing in light weather as her ancestor *Houdini,* maybe better!

"After getting all the pics we wanted, I nuzzled up behind, Charlie picked up the towrope, and I shut the Suzuki down and climbed on board. It was a very peaceful sail back to the marina, only just making way in the fading breeze, and as we silently slipped up the channel we came across a group of five Little Blue Penguins, natives of the very far south of our globe, something that very few people outside New Zealand ever see, and who were close to the northern limit of their natural range. A lovely sight, and perhaps a welcome to a little ship designed to go a long, long way further south as she was to head for Cape Horn later in her career."

Little Blue Penguins

Little Blue Penguins (LBP) are sometimes called Fairy Penguins, and they are the smallest penguins in the world. An adult LBP will tip the scales at a mere 2.6 lbs. The ones we saw swimming in Tauranga Harbor as we sailed back toward the marina looked quite delicate and each one had highlights of blue, which is apparently unique among all species of penguin.

The blue plumage runs down their backs, wings, and tail, and they have gray across their cheeks. Their short little bills are dark, black in fact.

The LBPs have quite large colonies in New Zealand, mostly in Oamaru, Chatham Island, and Banks Peninsula, but are seen all along the coastlines of New Zealand. We were lucky to get to see them within Tauranga Harbor. They say LBPs sometimes show up on the coasts of Chile, but these birds are called "lost." Who knows if they went there on purpose or not?

Another funny thing. We saw the little fellers in broad daylight, even though they are supposed to be nocturnal, moving about at night in groups of about 10 individuals. Apparently LBPs can dive as deep as 230 feet and stay underwater for about 35 seconds at a time. Most of their dives, however, range from 60 to 100 feet. Also, they say LBPs are noisy, but those we saw didn't make much sound. At least, I don't remember them making sounds.

LBPs mature at two years of age. They usually mate for life, although there are some cases of "divorce." The female lays two eggs at a time, actually one and then a second a day or so later. The male decides where the nest will be and usually chooses a well-hidden site.

The little penguins have a tough life. Eggs and chicks fall prey to gulls and skuas and sheathbills in the natural world, and even adult LBPs fall to rats, cats, dogs, and other non-native predators. Inanimate objects also present considerable danger to the little birds. Forgotten fishing nets. Plastic pollution. Reduction in food supplies (mainly anchovies and sardines, but also squid, krill, and small octopi) mean the little birds must travel farther afield for dinner, expending more energy than normal and traveling in dangerous waters.

At the time, John and I felt most fortunate to cross trails with a squadron of these majestic little birds, but then, a man often sees wonderful things when offshore in a little boat.

When Dolphins Come by Night

Back in the 20th century, I bought a Westsail 32 and christened her *DoriKam*, a portmanteau used in Japan for "Dreams Come True." I registered her in Honolulu, which was my last stateside domicile. Late in 1999, three of us sailed *DoriKam* from Olympia, Washington, up Puget Sound, and out the Strait of Juan de Fuca, crossing the bar into the Pacific Ocean.

We followed the West Coast, stopping in Portland, Oregon, and points south on our way to San Diego, California.

Why San Diego?

Because I didn't have time to make the trip all the way to Pichilingue, Mexico, which was my final destination. So I made a deal with Chula Vista Yacht Club for *DoriKam* to stay there until Christmastime, when I would sail her down and around the Baja California peninsula to Pichilingue. Once there, *DoriKam* would wait for me at the docks of Hotel Cantamar, which was owned and operated by my friend Fernando Aguilar. From there, I planned to explore the Sea of Cortez.

I arrived at Chula Vista Yacht Club from Japan in late December 1999, and spent a day waiting for the crew to arrive and for the oil in the diesel to get changed. After Fernando's son Pedro arrived, we bid the yacht clubbers adieu and steamed up San Diego Bay toward its mouth, 12 miles away.

Outside the bay, we set a course that would take us slightly east and south of the Coronado Islands on our way to Ensenada, the port of entry to Mexico.

Darkness fell when the Coronados were in sight off to the southwest. We motored on, as my schedule pressed and the wind was virtually nonexistent. *DoriKam's* propeller stirred up phosphorescence in the water and our wake glowed.

Then the dolphins came.

Sailing, the *chuff* of dolphins breathing often comes before you can even see the beautiful creatures. Motoring, the sound of the engine was louder than any dolphin *chuff*. So we were not aware of the dolphins until the pod surrounded *DoriKam*, playing in the bow wave and keeping pace on either side of the boat.

But it was dark, you might say. Dark. So how could you see dark dolphins in the dark water around the boat?

We could see them because they glowed.

But dolphins don't glow, you say.

These dolphins glowed. The only explanation I can think of is the same phosphorescent algae that caused the wake to glow was sticking to the dolphins' skin, letting off its eerie glow as they moved through the water.

We were speechless and mesmerized. Probably a dozen glowing dolphins cavorted around *DoriKam* for several minutes. I felt like I watched ghosts dancing around my boat.

The dolphins left as suddenly as they appeared, but the image of glowing dolphins will never disappear from my memory.

Aboard *Resolution*, sailing solo around the world, I hoped to add volumes of such encounters to my memory. After all, when I shuffle off this mortal coil, the only things I can take with me are what is in my heart and what is in my mind.

"Words are power. And a book is full of words . . . ."
--Yoko Ono

Chapter Sixteen

Prepare, Prepare, and Then Prepare Some More

The boat was finished (although there are always things that need adjusting or fixing). Only a few final preparations and an offshore shakedown before leaving for Hawaii, which was now my initial destination.

Why Hawaii?

I'd planned to sail across the Southern Ocean along the 40th parallel, dipping down to the 50s to round Cape Horn and back up to the 40s to make a landing at Stanley in the Falklands. But preparations had taken me into April and a winter voyage (remember we're talking about the Southern Hemisphere) around Cape Horn is not a wise idea. So I changed my plan, as I had changed it so many times. Remember? I was going to have *Resolution* in the water within a year of putting together her jig.

Going to Hawaii would take me north into the temperate zone. Further, a voyage from New Zealand to Hawaii would be an excellent tune-up for the circumnavigation. Plus I could stay in Hawaii until just the right time to sail around the Horn in the most placid season. Such were my plans.

But first John and I needed to take an offshore jaunt aimed at testing *Resolution*'s every system, from light wind sail combinations to heaving to when conditions got too dodgy for ordinary sailing. We figured to go at least 100 miles offshore from the Bay of Plenty's sandy beaches.

Tauranga's a port town that serves captains and crews of ocean-going vessels, so when it comes to outfitting for offshore work, the most economical method is to purchase from commercial chandleries that cater to the pros. That's where I bought my spare set of foul weather gear, a red suit that cost less than half what I would have paid at a yacht supply store.

That's the set of foulies I had on when I left my *Resolution* behind to be chewed into little pieces by the rocky fangs of Great Barrier Island. Those red China-made foulies hang on the wall of my bedroom as I write this account.

But let's go back to the offshore shakedown cruise.

The weather was snarky and the wind blew at gale force from offshore. Still, rather than sit the blustering wind out in cozy Tauranga Bridge Marina, we deliberately took *Resolution* out into the maelstrom that is the Bay of Plenty when the wind pipes up.

Still, the wind blew lightly as we left the marina. Yachts dotted the placid harbor waters, light weather sails aloft, but even the light wide ones built for speed around the buoys, did not . . . no, could not run away and leave *Resolution* wallowing in their wakes. That Energy Yellow boat was quick, light-footed, and gave immediate positive feedback to the tiller, which I held lightly with two fingers of a hand that had only three.

What a nimble, responsive boat! What a joy to navigate. What a precious adventure awaited her turnover from John to me.

Resolution was the first of the Sundowner class. John even called her the prototype. From his eyes, I suppose, as the first ship from the 40-odd sheets of plans was *Resolution*, indeed she was the prototype. And to her very last breath, she lived up to John's expectations.

Here's what John wrote about the Sundowner prototype.

"Sundowner ended up at 6.5m, (21 ft 4in) by 2.8m (9 ft 2in) on deck. She's a big boat for this length, full headroom, settees that make up two very comfortable bunks in the main cabin, space for a double bunk forward so she can accommodate four adults if that's the aim, space for a portable or marine head, chart table, a huge amount of storage, and an inboard diesel with 60 liters (16 US gallons) of diesel tankage. When she left, the prototype had over 200 paperback books, 300 liters (80 US gallons) of water and almost six months' worth of meals on board, in addition to all of the other stores and maintenance equipment for a long, long while at sea.

"Out on deck, the superbly sheltered cockpit has lots of lockers that include a dedicated paint and fuel locker that drains directly overboard for safety, a cargo hold under the after end of the cockpit floor, and huge scuppers that open out through the transom. The side decks are wide enough to take size 10 feet across the beam and the foredeck is positively huge for a small boat.

"She is styled like a classic working fishing boat. The long keel gives directional stability and reduces her roll rate, a big transom-hung rudder with a "wing" on the tip cuts pitching and increases efficiency, a hugely strong gaff rig with its mast in a tabernacle allows the mast to be dropped or raised without a crane, the cutter rig with its two headsails and topsail above the big main makes her a champion ghoster in light airs, and she can be shortened right down almost to storm canvas without having to change sails, which makes shorthanded sailing much easier.

"Designed for the amateur builder, everything in the boat can be built at home, using normal hand tools and materials from the local lumberyard. There is nothing here that is difficult, or beyond an average handyman. The same goes for repair and maintenance; it's quite conceivable that every tool used to build the boat could be carried on board when away voyaging. My personal philosophy for long range cruisers is 'if you can't fix it with what's on board, or do without it, then it does not come on board,' and with a boat like this, keeping to that philosophy is really easy.

"The prototype took a little more than a couple of years of work for a basically inexperienced builder to complete, and I had great fun helping on occasions, going to the launching, assisting set-up of the enormously strong rig, and fitting out of deck and running gear. Then I sailed with the skipper during the sea trials.

"She proved to be faster than expected. She felt very good to sail. And she tacked and steered very positively, pointed well, and was very well balanced, so it required little effort to keep her on course. We did not expect to keep up with the local coastal racers but over a few hours she was often still with the lighter and so-called faster boats. But her motion was much more comfortable than those lightweights, and the first real sail outside the harbor showed this up very well."

What's It Like?

Everyone at Funabashi Boat Park, where my Energy Yellow boat *Endeavor* is berthed, knows that I once set out on a solo circumnavigation. When I'm working on *Endeavor*, they come around to talk, and eventually the conversation always comes around to "What was it like, all alone out there?"

With *Resolution*, being out there was a study in confidence. I knew every bolt and every screw in her body. I knew every line and every stay in her rigging. I knew that she'd turn just as I directed her to, and that she'd stick to a course like she was on rails.

I need to say something about *Resolution*'s course-keeping capabilities. Bob McDavitt, weatherman extraordinaire, gave me the coordinates with which to set my course from Tauranga Harbor on the Bay of Plenty to some 50 miles off the western coast of Fiji island. The course was almost due north, 0 degrees, that is, but the wind made *Resolution* take a heading that fluctuated between 355 degrees and 345 degrees. Watching the compass, I could see that *Resolution* was throwing her head back and forth between those headings, and a look at the tiller, which was tied off so the handle was in the middle of the cockpit, as it would be if I had been steering by hand, showed me that it moved back and forth, stretching against the shock cord that supposedly kept the tiller up tight against the amidships tie off. But it always came back, and after a day of sailing with the tiller tied off as described, my Garmin told me *Resolution* had moved a few miles off the course set to Bob McDavitt's coordinates. Those miles of westward movement meant little in 24 hours. I decided that I would tack *Resolution* when she was 50 miles off course.

After about 36 hours of moving northward with the tiller tied amidships, the wind piped up even more, and I had to take hold of the tiller to maintain the boat's heading. *Resolution* was a little boat, but she sailed surely in the direction the captain pointed her in, and often sailed serenely with no one's hand on the tiller. Yes, she had a wind vane, but I had yet to hook it up.

Back to the shakedown cruise with John Welsford. He wrote:

"The coast off Tauranga is relatively shallow for many miles out, and there is nothing out there to break the waves that set up on the sandy bottom not far under the surface. It had been blowing gale force onshore for days, kicking up a monster swell from the southeast. Then the wind swung west south west at 20/25 knots, which made a 2m high chop running at right angles to the established 3m high storm surge. The combination made for an irregular and uncomfortable mix that had some waves combine to be a good 5m high.

"As a first real sail it was a great test of a new design. We sailed out to Mayor Island about 25 miles offshore, and carried on past until the island was very small on the horizon behind. We turned to port and headed across to the coast well north of the harbor entrance and sailed back down the coast. We covered about 80 miles in 24 hours, experimenting with the rig and the steering all the way and found that the boat could be made to self-steer on most points of sail with just a shock cord on the tiller. We tried her hove to under different combinations of sail, and that made a tremendous difference to the motion. We also found that we could cook down below when the conditions outside were a lot like the inside of a washing machine and that the cockpit was a very secure and comfortable place for us to weather a storm.

"Our little cutter proved to be a very capable sea boat, easy to handle, not at all sensitive to trim or conditions, easy to work, faster than most would expect, and very capable of making long voyages in reasonable comfort. We took turns on watch in the deep and comfortable cockpit, did a couple of changes of sail, finding that the combination of wide side decks and cutter rig made for an easily worked rig. Off watch, we slept surprisingly comfortably in the big bunks down below considering the horrific conditions, and although it was not the sort of weather I'd choose to go out in, it was a very good experience to look back on. We had a good time and while the little boat's first sunset was a very black and threatening one, her first dawn at sea was just beautiful.

"Before I handed the boat over to the owner, we sailed in every condition from flat calm to full storm. She was designed from the first to be as self-sufficient as possible, and to go long distances shorthanded. I did many miles in the prototype during her sea trials and I'm as pleased with the Sundowner design as I have been with anything I've ever drawn."

Sights on the Sun

One of the pieces of equipment that went down with *Resolution* was my sextant. A decade or more ago, a captain could not put his entire trust in electronics (nor should a captain do so today, either). A saltwater environment is not and will never be a favorable one, even for the most sophisticated of electronics.

So I set about to learn how to calculate my approximate position from the angle of the sun to the horizon at noon. I never did learn star shots, but then, I never needed a sun shot either.

I drove my tired Mitsubishi up north of Auckland once a week to take a lesson in navigation from Barry Young. I'd do more than one sight at a time (I was brushing up what I'd learned when I obtained my 1st Class Boat Operator's License in Japan, which qualifies me to operate any vessel of up to 20 tons anywhere in the world), just before, at, and just after noon. I actually got pretty good at it, but don't know if I could do it now without a refresher.

Barry taught me then, and now he teaches navigation to fishermen in the Pacific Islands. He was also the volunteer master of the square-rigged training ship *Spirit of Adventure* at one time.

I bought my bronze sextant in its wooden box from TradeMe, and learned how to set the mirrors and sight the sun at noon from Barry. But the sextant was my fallback. I did my navigation with a Garmin 72 and paper charts, which helped me set a point on the map at which to aim *Resolution*. As I said, the point, west of Fiji, was provided via iridium satellite telephone from Bob McDavitt, the weather guy who kept an eye on my voyage.

I downloaded paper charts from the internet and printed them out full size. I also had a UHF radio with a whip antenna attached to the pushpit rail, and a GME AccuSAT emergency beacon, just in case. So, in my mind's eye, I was basically ready for anything, navigation wise. If I had only known. If I had only surmised. If I had only . . . .

Our rough little shakedown sail showed how well *Resolution* handled storm-strong winds, steep waves going athwart the Pacific groundswell as it surfaced on the shallow bottom of the Bay of Plenty. It showed how well she pointed into the wind, even though she was a most buxom lass. It showed how well she rode the waves and how quickly her roll subsided after a wave went by. *Resolution* was a sea boat. All she needed was a sea captain.

Starbucks on The Strand

I sit in Starbucks at the Aeon Mall, which is not far from my home in the city of Chiba, Japan. I come here out of habit, because it is my writing place.

When I arrive a few minutes before noon, all I have to say is, "Lunch," and the baristas pop my ham and cheese sandwich in the microwave and fix my favorite drink. The Strand is a street that runs along the edge of Tauranga Harbor, putting my New Zealand Starbucks a six- or seven-minute drive from the yacht harbor in my Mitsubishi. Starbucks fronts The Strand. As in Japan and Ohio and Arizona, I sought out Starbucks as my first choice of places to write. Most of my manuscripts are first written by hand and no barista pushes me to leave my seat just because I've been there an hour or so. Yes. Starbucks is a good place to write.

John and I drove from Tauranga Bridge Marina to Mt. Maunganui, which was once an independent township. From there, we located The Strand and followed it to the roundabout where the street ends. Starbucks stands on the corner closest to the marina, complete with typical Starbucks sidewalk tables. The seats where I quaffed Signature hot chocolate as I created fascinating prose had changed character. Or perhaps my character had changed. John and I took our beverages to a sidewalk table where we enjoyed the balm of the afternoon and watch bicyclers and skimpily dressed young women, commenting on neither. We took our time, as Starbucks tends to encourage a person to do.

At Last, Departure

The time came at last. All my checkouts with New Zealand immigration were finished. All *Resolution*'s equipment was checked and double-checked. And all the provisions and water necessary for the leg from Tauranga to Honolulu were loaded and stored, properly I hoped.

Resolution was registered in the state of Rhode Island, USA. Why in the world would I register my circumnavigation vessel there? I'll tell you. Back in 1632, John Whipple arrived in Massachusetts in the New World from England. When he finished his indentured servant contract, he moved to Providence Plantation in Providence, Rhode Island. His home there still exists.

Fast forward to 1733. That's the year Abraham Whipple was born in Providence. They say Abraham didn't get much of a formal education and most of his acuity came from a naturally curious mind and began with his signing on as a cabin boy aboard a vessel trading with the West Indies. Soon he was commanding his own vessel, and then he captained a privateer. Some say he made more than a million pounds for the merchants who financed his privateering.

As relations with Great Britain declined, Abraham Whipple's reputation soared. He and a crew from the local pub burned *HMS Gaspee*, a customs cutter that went aground when chasing the packet ship *Hannah*. The place the ship went aground is now called Gaspee Point, and Abraham Whipple made a name for himself.

The Continental Congress commissioned four man-o-war vessels, one of which was put under the command of now Commodore Abraham Whipple. With this vessel, he fired the first cannon shot at the British from a Continental Navy warship.

So you see, I have salt water in my blood, and it comes from Providence, Rhode Island. No wonder, then, that I registered *Resolution* in Providence.

June 15th, 2008. Two years and 260+ days from the day I landed in New Zealand, I was set to leave in *Resolution*, a 21-foot gaff cutter of my own and myriad helpers' construction. Sails by Tony Thornburrow. Bronze work by Silverstream Foundry. Halyards and lines and all manner of yacht ropery from Quality Equipment Ltd. Bronze keel bolts, etc., from Port Townsend, Washington. Yacht stuff from Burnsco. Energy Yellow paint from Home Depot. Lots of carrot cake and delicious vittles from Fran's Cafe. Gallons of Signature Chocolate at Starbucks and heaps of bonhomie at WOS, even though I was not a "naturalist."

Funny. I don't remember a birthday party from when I was a lad in the mountain town of Show Low, Arizona. Surely I had some, but as my birthday is the day after Halloween, maybe the memories are twisted together.

Following graduation as a BA in Asian Studies, I had six months before I was to be in Hawaii to do post-graduate work at the East-West Center, which was a school set up within the University of Hawaii. During that time back in Show Low, I worked in a paper mill 26 miles away . . . which has nothing to do with the story.

As resident expert on Japan, I was once asked to give a talk to some people gathered at the local LDS church. Such evening meetings were called "Fireside Chats." In particular, I remember that in the Q&A segment after my talk, a local man spoke up with a question. "Say. Tell me. Do they got a Fourth a July over there in Japan?"

Of course I knew he meant Independence Day, but I couldn't resist saying, "Sure."

"When is it?"

"Between July 3rd and July 5th," said I, hoping to bring the house down.

Not many people laughed.

Half the Trip Is In the Going

Sunday, June 15th, 2008. The day *Resolution* should have set sail for Honolulu. But Bob McDavitt, weather guru at New Zealand Metservice, suggested I put off the departure for a couple of days. Better weather then, he said.

I notified all with an interest in seeing *Resolution* off and continued work on the boat. What is it they say? A boat is forever a work in progress.

Monday it rained.

Tuesday it rained.

That meant Wednesday, June 18th, was D-Day.

I started filling 2-liter plastic milk bottles at nine in the morning. Ten o'clock saw nearly 100 liters of water stored in forward lockers and under the hanging locker amidships.

But where was I to put the remaining 200 liters? A 20-liter jerry can in the forward compartment. Two 20s on top of the starboard locker. Five 10-liter jugs under the galley for 110 liters. Five 3-liter bottles under the Porta-Potti base. Two 20-liter collapsibles on the floor of the forward compartment. Three more 3-liter bottles forward of them, and a 5-liter one with a tap in it strapped in the galley. In all, *Resolution* carried 279 liters of water for a maximum 100-day voyage.

Stores for the voyage fit neatly beneath the two berths in the main cabin. A hundred and fifty portions of freeze-dried rice. Fifty portions of cream stew and 50 of curry in foil retort pouches. A meal of rice and curry, for example, meant only opening a packet of rice, pouring a cupful of hot water inside and re-sealing it for half an hour. The foil pack of curry went into the remaining hot water on the stove to heat while the rice reconstituted. In half an hour, steaming rice in a bowl with delicious golden-brown curry poured over it. Warm, and easy to whip up.

On the starboard side, 85 250ml cartons of UHT milk kept company with about the same number of orange-mango juice cartons. Nearly 300 muesli bars filled another locker, accompanied by 5 kg of peanuts and raisins.

Other stores included two large sacks of rolled oats and 1 kg of raisins to put in the porridge, 3 kg of scone (biscuit in the States) mix for a taste of fried bread along the way, a large bag of muesli to go with the milk at breakfast, and two large cakes baked by friends Ian and Karen from Cambridge.

I didn't forget fresh food. In nets along the port side hung a dozen oranges, fifteen lemons, ten limes, two hands of green bananas and one nearly ripe, and a bag of twenty or so carrots. Plus, John Welsford brought along two large grocery bags of apples from Newstead Orchards, where I bought apples at least once a week, in season, all the while I built *Resolution*.

People began arriving about noon. Bruce, Tom and Ann, and James and Kobie from Hamilton. John, *Resolution*'s designer, and his wife Denny, also from Hamilton. Blair Cliffe had come down the day before, assuming we were going to leave on schedule, rain or shine. P.J. the truck driver, came over on Sunday with his wife and two sons. What's more, with an article on the planned voyage in the *Bay of Plenty Times*, interested readers stopped by berth F21 where *Resolution* was moored every day, so I spent almost as much time talking as I did getting her ready for sea.

After lunch, John got things organized in the cockpit. Bruce and James hanked cordage: 50-meter No.2 anchor rode, all the 8mm and 10mm yacht braid I had on spools, and miscellaneous other lines I'd left in the cockpit. By the time stores and water were stowed, *Resolution* had become an organized boat.

Customs arrived at one, Kevin, the main man, with a gaggle of four junior officers tagging along behind. Formalities were over in a few minutes, and the group left *Resolution* with a warning: We don't want you coming back to New Zealand . . . unless you absolutely have to.

The whirlwind of preparation continued.

I told Kevin we planned to leave about 4 p.m., but it was five before I started *Resolution*'s 7.5HP Bukh powerplant.

I thumbed the mike on the VHF. "Bridge Marina. Bridge Marina. This is *Resolution*. Preparing to leave berth F21. Thanks for everything."

Fred's voice came back over the wireless. "*Resolution. Resolution.* This is Bridge Marina. Been good having you here. We'll give you a blast on the way out."

Hugs and handshakes. A wet eye or two. Mooring lines cast off and coiled in the cockpit. At 5:10, I put the morse in reverse, and *Resolution* slowly backed from her pleasant home of nearly two months. Friends waved from the marina's pontoons. I shoved the morse into forward and pushed *Resolution*'s tiller to starboard. Obediently, she carved a smooth curve in the quiet water, turned her bow to the setting sun, and moved toward the mouth of the marina's breakwater. As we came abreast of the marina offices, a horn blasted and marina employees waved goodbye to *Resolution*.

"If you are brave enough to say goodbye, life will reward you with a new hello."
--Paulo Coelho

Chapter Seventeen

Honolulu, Here We Come

March 2018: From Bridge Marina, we took the street that runs parallel to the Tauranga Harbor Causeway all the way to the foot of Mauao, the ancient lava dome better known as The Mount, which protrudes from a sandbar that connects it to the mainland. There, John and Indy let me off at the entrance to the path that goes around the dome. Walkers, arms swinging rhythmically, joggers, choosing whatever pace suited their physical condition, and Chinese tourists, obviously on their first trip to New Zealand, populated the trail. John and I walked, but not with the rhythmic pace of the exercisers.

We paused frequently, usually to gaze at the clear aqua water that surrounded the mount, and to appreciate the collection of pleasure boats anchored just off the harborside beach.

John stopped where he and the WOS well-wishers had waited for *Resolution* to pass on her way north, headed for Hawaii. "You came through the channel pretty as you please," he said. Kauri trees with overhanging branches framed the picture and I remember how my WOS friends hollered their goodbyes and bon voyages and flashed their handheld torches at me. At the moment, I had waved back and felt good and safe.

June 18th, 2008: Tauranga Harbor lies behind The Mount and connects with estuaries that spread south and west. Tidal currents race in and out of the harbor entrance at three to four knots. We left shortly before high tide and didn't have to push against the current as we made our way down channel. The sun disappeared, leaving only the gunmetal gray of dusk and the dark waters of the harbor. Marker buoys flashed green to port, showing where deep water left off and shoals began.

Late autumn in the Southern Hemisphere. South winds blow cold and northerlies bring warm air from the subtropics. The two days of rain that delayed our departure came from the north. Even overcast, temps were in the high teens, 16 degrees – 17 degrees Celsius. But Metservice predicted winds from the southeast, and that would mean drops in temperatures at night. At the dock, I'd dressed for chilly nights. My inner layer was merino wool from Icebreaker, covered by my normal trousers and T-shirt. Yellow and navy blue Gill foul weather gear, just the trousers at first, went on last to keep the cold southerlies at bay.

As *Resolution* rounded The Mount, I was toasty warm. Flashes rippled on the side of The Mount as we motored out of the channel. Shouts drifted across the water. The well-wishers from the marina had driven to Mauao, parked in the lot, and walked around the footpath so they could give *Resolution* one last hurrah in final send-off. I raised an arm in answer and afterwards learned that John and the others on the path were able to see my wave silhouetted against the remains of the day.

The wind whistling around The Mount turned cool, so I retrieved my beloved dark-green fiber-filled jacket from below and put it on.

John and I decided at the marina that *Resolution* should fly her storm trisail instead of the main until I could judge the winds better and *Resolution* and I were farther north into the South Pacific highs. We raised the trisail at the dock and *Resolution* motored all the way to A Buoy at the harbor entrance with its help. Past the A Buoy, I raised the staysail and unfurled the jib. *Resolution* barreled along at about five knots on a heading of 020 degrees, just east of north. Our first waypoint, assigned by Bob McDavitt at Metservice, was S20 E179.

The wind picked up and spits of cold spray came over the cockpit coaming. The old saw goes: If you wonder whether it's time to take down a sail, it is. I furled the jib and *Resolution* sailed on, still making 4.5-5 knots. Over the shallow bottom of the Bay of Plenty, the swell often exceeds 2m and stiff winds build sharp waves that frequently move crossways to the swell. The seas were as rough as those we encountered on the shakedown cruise around Mayor Island, maybe rougher. I settled back to steer *Resolution* through the night as the wind vane oar was tied up out of the water and I had no autopilot. I had no choice but to keep one hand on the tiller and one eye on the red of the LED-lit compass. Overhead, the tricolor navigation light oscillated wildly atop the mast, some 10m above the ocean surface.

Although the wind was sharp, the sky was clear and blue-black, except for a few clouds that obscured the moon when it rose. Visibility must have been about 20 miles, because I could see Mayor Island well. At first, it was a dark spot on the horizon at 10 o'clock. But gradually we pulled abreast of the island, sailed down its rugged eastern coastline, and then left it behind as the moon crossed the sky.

June 19th dawned with a cloud cover. The wind blew briskly, but we made only 3.5 knots under Storm Orange staysail and trisail. *Resolution* sailed on her bottom and it felt like we'd passed over the Bay of Plenty shelf because the swell had smoothed out and the waves seemed more regular. I let out the jib and sheeted it tight to port. The wind came from east of northeast, which made it difficult for *Resolution* to point as high as 020 degrees. She settled in at due north.

I'd steered all night and wanted some rest. Before leaving, I prepared a tail with a block on it and another with two lengths of rubber hose lashed to a loop spliced into its end and to two grommets made from 10mm polyprop line. The tail with the rubber tubing was coiled in the canvas bucket, but I couldn't find the one with the block. Rather than rig a jib sheet through a block to the tiller, I merely ran a line from a cleat on the lee winch pad to the tiller. I set its length to where the weather helm began to pull on the rudder. From a corresponding cleat on the windward side, I ran the tail with the rubber tubing, which I call the "spring tail," to the tiller. The grommets slipped over the tiller and came up hard against a 3/8-inch bronze rod through the stick about a foot from its forward end. To my relief, *Resolution* held her course like a train on rails. I was free to do something besides steer.

I dug the Garmin GPS unit from its lair deep in a coaming locker. I don't remember the exact position, but we were barreling northward at 4.5-5.0 knots and were less than two miles west of the line to our waypoint. I could live with that.

In the notebook that served as *Resolution*'s deck log, I entered the position, the barometer reading, and a couple of comments.

For the first time since getting a couple of muesli bars just after midnight, I went below.

Chaos.

The wild ride over the shelf of the Bay of Plenty had exacted a toll. Nets that were hung on plastic hooks and filled with fresh food to port and clothing to starboard were now mostly on the cabin sole. The swinging motion of the heavy waves made the nets chafe through at the hooks.

Stepping down into the cabin, I found it slick and wet. Liquid slopped between the floors at the galley, in the main cabin, and in the forward cabin. And there was a tang like kerosene in the air. Not such a good idea to recycle milk bottles as water containers, I thought, remembering the two-turn caps on those plastic bottles. Obviously, some had sprung leaks. My hurricane lantern hung from the ceiling, belayed by a length of shock cord that kept it from hitting anything as it swung. I thought the lantern was the source of the slick and the smell, that some of its kerosene had dribbled onto the floor. I went to work with bucket and sponge. In an hour, the place was presentable again. I laid my blue and green Hawaii beach towel on the floor to sop up the last of the film of slick stuff. The cabin sole was no longer slippery, and a quick check of water bottles showed that most of the milk containers were holding up. *Resolution* sailed on, oblivious of my cleaning and straightening up.

June 19th. The kind of day every voyager dreams of. Not once did I have to adjust the tiller and GPS checks showed *Resolution* edging west but still within five miles of the line to our waypoint. Birds soared over the waves. They had pale bodies and blackish scimitar-like wings with a pointed joint. Mollymawks, perhaps. I spent more time than I should have, maybe, watching them swoop and wheel. But then, Honolulu was two months away. No rush. Sit back. Relax.

Checking the compass, I saw that the wind had come around slightly, bringing our course west of due north. We were now just over five miles west of the plumb line. Still, I could live with that. The South Pacific chart I was using showed me more than 100 miles offshore, and our course would not bring us uncomfortably close to land. I furled the jib for the night and allowed *Resolution* to take her northerly course while I got some sleep—after eating a red Braeburn apple, two bananas, and an orange. I tossed the banana and orange peels overboard. Bio-degradable.

Before leaving Tauranga, I re-read John Letcher's book on self-steering for sailboats. In it, he charted how the chances of colliding with a steamer were tiny and sleeping for an hour or two at a stretch posed an infinitely small chance of danger. I agreed with him, and settled in for some winks.

I woke three times for looks around the horizon before getting up at 6 a.m. The gray of early morning was just starting to spread in the east. Breakfast consisted of an apple and a banana, which I ate in the cockpit while the lashed and sprung rudder steered *Resolution*.

June 20th. I took our position from the GPS. On the chart, the mark showed us farther north, perhaps 150 miles from Tauranga and, according to the Garmin 72, still only slightly over five miles west of the waypoint line. No problem.

I drank a couple of swallows from the 3-liter water bottle I kept in the cockpit, then remembered the lights and reached into the companionway to turn the battery switch to OFF. "Must switch to kerosene navigation lights," I said aloud. Must save battery power.

The wind shifted, and the best *Resolution* could do towards the waypoint was 350 degrees. Our distance to the west of the waypoint line increased steadily. I decided I'd tack and head east when we were 50 miles west of the line. Until then, barring a mighty wind shift, we'd plow on, heading north-northwest. I was more eager to make northerly progress than easterly. In retrospect, perhaps I should have tacked and gone east right then.

With just the trisail and the staysail, and the wind at about 20 knots, we were making only about 3.5 knots through the water. Time to unfurl the jib; an easy job. I'd not sailed with a furling jib in the past. Don't think I'll sail without one in the future.

With jib unfurled, speed jumped to 4.5-5 knots, and the compass rose settled at about 330 degrees. I was moving northwest when I wanted to go north-north east, but still, I'd made my decision to stay with *Resolution* as she was until we were 50 miles west of the waypoint line.

Again, sailing *Resolution* was a joy. The storm trisail was a bit small for the power of the jib and staysail together, but adjusting the tiller to the balance point, lashing it to a cleat on the lee coaming, and springing it to a similar cleat on the windward coaming made the boat track constant to the wind.

I dug a couple of muesli bars from under the starboard bunk squab, along with a 250 ml carton of mango-orange juice. Breakfast was a mish-mash of roasted oats, honey, and the tarty-sweet taste of mangoes. It was good.

Apples and oranges and bananas and muesli bars constituted my diet for the first days of the Honolulu leg. And two days out of Tauranga, the time came to put the galvanized bucket in the cockpit to use.

Resolution had a head, a 20-liter Porta-Potti. And it was fully charged, too. But salty seafarers offshore use an oaken bucket . . . or in the case of *Resolution*, a galvanized steel one. Large cruisers, with space to burn, have seats into which the bucket slides after three or four inches of seawater has been poured into it. *Resolution*'s bucket fit neatly under the cockpit seat lip, which extended nearly four inches all around, with nicely rounded corners. So, if the bucket user bares himself (or herself, as the case may be), spreads his legs to approximately 60 degrees and sits on the cockpit seat overhang as if it were a toilet seat, his business end is located right over the galvanized bucket.

So I dragged a couple of quarts of water from the ocean with my handmade canvas bucket and poured it into the galvanized one. *Resolution* steered straight ahead as if she were looking the other way while I did my duty. Off with the foul weather jacket. Unclip the suspenders of the foul weather pants. Bare the essential part of the anatomy. Take the required seated position. Do duty. Clean away any residue with paper towels prefolded and torn in half. Once used, the towels went over the side.

The entire system worked perfectly. Once the job was done, the contents of the bucket got dumped over the side. The bucket then got washed out with the contents of another dip with the canvas one. Simple and clean.

I also found the perfect position for relieving myself. Kneeling on the leeward cockpit cushion (which was really a 75mm closed-foam fender) put me where I could lean my right shoulder on the wooden pad that held the 2HP Suzuki outboard, put my left against the Danforth anchor lashed to the pushpit, and hold on with my left hand while peeing into the brine. Life at sea. Later, however, someone asked why I didn't just use the bucket. I had no answer.

The wind piped up in the afternoon, and I rolled up the jib. We still made more than 4 knots. Apples for lunch. Oranges and bananas for tea. And *Resolution* plowed ahead on that unswerving heading of 330 degrees. We were making a lot of westway, but not yet enough to be 50 miles off the waypoint course.

The wind continued to rise, and now whipped mare's tails off the wave tops. Some spray would reach the cockpit and one wave sent bucketsful of cold, salty water down my neck and into the cockpit. The water immediately disappeared out the drains, but not from inside my oilies.

By midnight, when I made my all's well call, a storm blew and the rigging howled. I lowered the staysail and we labored on under storm trisail alone, the motion exceedingly rough. Still, all was well. Or so I thought.

A GPS reading and a chart plot put us nearly 20 miles west of our waypoint line, and about 80 miles off Northland on the eastern coast of New Zealand's North Island. I decided to heave to for the night. With the tiller tied about 15 degrees to leeward and the storm trisail sheeted hard to the port quarter, *Resolution* settled down and fore-reached at about 330 degrees. The GPS said our leeway was 2.6 knots. I could sleep for several hours and not go on the rocks of Northland.

Below, I found chaos again. And the floor was slippery wet with . . . fuel! It could only be fuel. A suspicious liquid slopped against the hardwood floor between the galley and the chart table. Something had to be wrong with the main fuel tank. How else could this much diesel end up on the main cabin sole? I swiped the Hawaii towel along the floor with my foot, but it was already soaked. Although *Resolution* was hove-to, she still bucked and I could hear a hollow pop as diesel splashed across the inlet hole in the fuel tank. What was it? Where would so much fuel leak out? I had no idea, but there was nothing I could do about it in the dark. I doubled the Hawaii towel lengthwise and laid it across the floor. Maybe I could get some sleep, as the diesel smell didn't seem to bother me.

An apple rolled along the bunk upright. It looked bright red and shiny in the light of my headlamp, but I knew the shine was petroleum. It would take soap and water to make it edible. Later.

I settled down in the leeward bunk with my legs bent so my feet were on the floor. Didn't want to get my sleeping bag and fleece blankets rank with diesel.

Finally, my throat feeling a bit raw from the diesel-laced atmosphere, I felt sleepy. *Resolution* rocked with the waves and the wind sang in her rigging. A pot sat on the gimballed stove in the galley, clanking against its retainers with every sharp wave. Still, I fell asleep amid the cacophony and the motion, and didn't wake until nearly dawn.

Wondering about leeward drift, I checked the GPS as soon as I got up. I used the plastic scale my navigation teacher Barry Young gave me to calculate the drift. Only a little over 15 miles. But the GPS now showed us nearly 30 miles west of the waypoint line.

The wind still blew at near storm strength, and I didn't want to raise the staysail just yet. So I decided to deploy my parachute anchor to cut the leeway drift to around half a knot.

That was the plan.

I donned a safety harness, then an inflatable Mae West. With my butt on the cabin top and feet on the side deck, I slid along the windward side until I could click the harness tether to one of the eyebolts at the base of the mast. Harnessed to the boat, I knelt on the foredeck and opened the portside locker at the forward end of the cabin. I'd stuffed a 12mm nylon sea anchor rode in the locker before leaving Tauranga. I pulled out the bitter end and tied it to the Samson post with a bowline. I flaked about 50 feet of the rode into the anchor well, clove-hitched the rode to the Sampson post again at that point, then flaked the rest of the rode into the well.

The bronze pad eye for the parachute anchor's snatch block was on the bowsprit, about six inches outboard of the stem. I ignored the bucking motion of the boat as I snapped the snatch block home to the pad eye, then snapped the rode through the block. I led the thimbled end of the rode over the whisker stay and back past the pulpit before threading it under the lifelines and over to the starboard deck locker.

The locker contained the parachute anchor, and its big stainless swivel hung from the thimble at the end of the shrouds. I pulled the anchor out of its locker, spread it on the deck against the locker, unbraided the shrouds, and shackled the swivel to the thimble in the rode. All was as practiced to that point.

The parachute anchor setup requires a bridle from the coaming winch to a snatch block that fits over the anchor rode. Herein lay my problem.

I had to launch the anchor, play out the rode, and affix the bridle, all at the same time.

Launching the parachute anchor was no problem. It went over the windward rail, just forward of amidships, slid down *Resolution*'s hull into the water, and began to play out. In retrospect, I should have snapped the bridle's snatch block onto the rode while it was still amidships. Instead, I clipped it on at the bow.

Too late.

Resolution had already turned bow to the anchor, and there was no leverage with which to make the bridle slide out to where it would hold *Resolution* at 60 degrees to the wind.

Larry Pardey talks of the problems of heaving the parachute anchor off the bow. Every 10 minutes or so, the boat crosses the rode and goes to the opposite tack.

At that point, *Resolution* was in no danger, as there were no high-breaking waves. Once in a while a shape like the volcano cone of Mt. Fuji would rise a few hundred yards away, but the cone flattened out long before it reached *Resolution*. Sitting on the forward hatch cover was a wild ride, but at no time did I feel *Resolution* was in danger, and the decks were basically dry. A bit of spray came aboard from time to time, but no solid water did.

Pardey gives instructions on how to get the bridle out if you miss the first chance. He says for one crewmember to let out 20-30 feet of rode while a second crewmember hauls in the bridle. The loose rode will allow the bridle's snatch block to ride out to where it can hold the boat in the proper position. Problem: I had no second crewmember. For me, Pardey's system wouldn't work.

I sat on the hatch cover and contemplated the problem. I tried loosening the bridle so the snatch block ran down the rode and into the sea. But the rode's angle was too far forward for the bridle to get any purchase.

I worked on the problem for more than an hour and never found a solution. All I can think of when single-handing is making sure the rode feeds through the bridle block from the outset. You only have one chance. After that, it's a matter of deciding to ride out the storm while tacking back and forth. The major problem is chafe, as the rode rides under the bobstay when the boat is on the opposite tack. After three or four tacks, I could already see broken threads where the rode rubbed on the bobstay.

Total frustration.

I thought my way through the situation. Who knew when the wind would die down, though it seemed to be lessening a bit. The cabin floor was slick as an ice rink and dangerous to navigate. A leaky fuel tank would put me in danger of running out long before reaching Honolulu. Should I push on? Should I turn back, go to Tauranga, fix the fuel problem, and start over? *Resolution* tacked again.

I unclipped the tether, slid back to the cockpit on my butt, and went below. Things were still a mess, as I had not cleaned up the night before. In the light of day, the level of liquid slopping at the floors looked higher and the level in the main fuel tank was definitely lower than it had been after I filled it at the pumps in Harbor Bridge Marina.

What were the options? How many times would I hit rough weather on the way to Honolulu? No way to know. Actually, the sail around Mayor Island was a rough as anything we'd come through since, and there had been no trace of fuel on the floor then. How much fuel would have to leak out before it stopped?

At 66 years of age and something over 100kg body weight, I was not in the shape of an Olympic gymnast. A slip in the cabin could end in a broken rib or worse. The solo circumnavigation was on no specific timetable, except to round the Horn in the dead of summer. That gave me a couple of months' leeway. Better to return to Tauranga to fix the fuel leak than risk unnecessary injury. I went back out on deck.

The wind had eased a bit more.

Out on the foredeck, tether clipped, I considered ways to retrieve the parachute anchor. With an index finger and part of a thumb missing from my left hand, gripping a wet rope for hand-over-hand retrieval was an iffy proposition. I'd not tried the anchor chain windlass with rope.

But I'd made the decision. Back to Tauranga.

"Drink from the well of yourself and begin again."
--Charles Bukowski

Chapter Eighteen

Back to Tauranga

Despairing retrieving the para-anchor, I finally cut the rode. I had another anchor on board and the No.2 anchor rode could serve as parachute anchor rode if necessary. And I told myself I'd buy another 12mm Dacron line when I got back to Tauranga.

I raised the staysail and prepared to sail toward Tauranga. The GPS put me in a position where a course of 180 degrees would allow me to skirt Great Barrier Island and the end of Coromandel Peninsula on my way back.

The wind was down to 20 knots or so. *Resolution* sailed sedately. As evening fell, I sent John Welsford a message on the sat-phone. RETURNING TO TAURANGA. SAFETY ISSUES. ONE MAJOR ONE MINOR.

The minor issue was a loose tiller pocket. Wave action causes the rudder to move back and forth against the tiller. A crack had developed between the starboard cheek and the rudder stock, and I was afraid it would eventually pull off. I dug in the starboard cockpit locker for a 3/8-inch stainless bolt with nut and penny washers. Then I got out the cordless drill and fitted it with a 9.5mm bit. I leaned over the quarterdeck and drilled a hole through both cheek pieces and the rudder stock. The drill had just enough juice to get the job done. The hole was on a slant, but it allowed me to get the bolt in, washers on each side, and clinched down tight with an adjustable crescent wrench that Kiwis call a spanner. It should hold to Tauranga where I planned to put in a line of bolts above and below the tiller to hold those cheek pieces on tight.

Ship of Dreams

Resolution maintained her heading of 180 degrees though the night, and the day allowed me to do some cleanup in the cabin. The fuel slick, however, remained at a dangerous level, and I took extreme care whenever I went below.

We sailed without incident, except for the sorrow of turning back.

June 22nd. Clouds roiled and squalls crossed our path several times during the day. *Resolution* took them in stride, good sea boat that she was. I sent a message to Chuck Lienweber, who was tracking me on the *Resolution* Around the World website. Turning back, I wrote.

Nothing to do now but steer a course of 180 degrees, and *Resolution* was doing that on her own.

The wind had been fluky all day, but constant in direction. It blew for hours at 20 knots, then jumped to 35 or more with huge dollops of rain, and then back to 20. I saw the squalls coming, but I kept *Resolution* in staysail and trisail, so she could handle anything up to full storm. The clouds piled up in the west. There was land over there, but not close enough for me to see, and I should've been able to see at least 20 miles from *Resolution*'s deck.

The course line on the chart showed us clearing Great Barrier Island and the tip of Coromandel Peninsula by what I considered a safe margin. I made my midnight call. As I put the sat-phone back in its deep shelf on the cabin bulkhead, lightning flickered in the west. Probably another squall, I thought. I took a quick 360 degree scan of the horizon and saw no looming shapes, ship or shore.

The Way it Was

As I stood in the companionway and did that 360 degree of the horizon, a white light flashed in the periphery of my vision. I snapped around and peered west in the direction the light came from. Warning flasher? Land lay to my west. The rocky coastlines and rugged islands that guarded the entrance to Hauraki Gulf.

No regular flash. A moment later lightning flickered, followed by a distant crackle of thunder. My chart of the South Pacific showed me with sea room to starboard, enough to pass safely east of the rocks as *Resolution* and I returned to

Tauranga Harbor at about 4.5 knots. The squall with its flickering lightning and booming thunder hit us with a blast of wind and the rushing sound of rain on waves. Waves had been roaring at us all night, pushed by 25- and 30-knot winds out of the northeast. But *Resolution* had proved her worthiness in windy situations and the squall only served to quicken her speed and belly out her storm orange trisail and staysail a bit more. Minutes later, *Resolution* settled back to her course of plus-minus 180 degrees.

As strong as she was, *Resolution* could not continue the Honolulu leg of our solo circumnavigation. Something was wrong with her fuel tank. In heavy weather, diesel leaked somewhere and found its way somehow to the main cabin sole, which made standing on it life-threatening. I felt like ordering a pair of ice skates.

Reluctantly, after a night hove-to in storm conditions, Captain Me made the decision to return to Tauranga to find and fix whatever was wrong. At that point, I thought it might mean replacing the fuel tank. We wore ship and headed south. I wanted to head south anyway, but *Resolution* liked a heading of 210 degrees, and kept slipping to that point, 30 degrees west of south. Working with the set of the sails and the rudder balance, I got her to vacillate between 170 degrees and 200 degrees.

Once *Resolution*'s sails were balanced on a course and the tiller lashed in place, she would steer herself as long as the wind stayed constant.

We wanted to go to Honolulu, but the oil-slick cabin floor was just too dangerous for a captain far past his prime in terms of strength to body weight. So we sailed southward through the squall, confident of sea room on our way outside Hauraki Gulf and down the coast of Coromandel to the quiet harbor behind Mt. Maunganui at Tauranga.

I'd called home on the Iridium sat-phone at midnight. All is well, I'd said. The squall passed, in all its sound and fury, and all was well. Swift calculations from the Garmin 72 GPS seemed to put us a good way east of Great Barrier Island. I checked the tiller setting, the compass course, and the balance of the staysail and trisail. All was well. *Resolution* would sail forever, constant to the wind. I couldn't keep the smile from my face. No skipper could have greater confidence in the ability of his ship than I. Below, I ate three muesli bars and had a bracing drink of water from one of the 2-liter milk bottles I'd recycled as water containers. I sat back against the pillows that lined the curve of the hull on the lee side. *Resolution* heeled obligingly at about 15 degrees to make the seat angle as comfortable as any recliner in any landlubber's rec room.

The Sandman came. I shook myself and took one more 360 degree gander from the companionway stairs. All was well.

My feet were dry. My clothes were dry, as I'd changed into my second set of foul weather gear and hung out my Gills. I slipped off my Sperrys and burrowed under the unzipped sleeping bag to snatch some winks.

Bam.

Bam.

Thunk.

Strange wave patterns, I thought through the sleep haze. But waves sometimes thunked against *Resolution*'s hull when squalls came through or when an odd wave hit.

Bam. Bam. Bam.

I threw off the sleeping bag and slipped on my Sperrys. Hauling the hatch back and stepping up onto the first companionway step, I expected to see towering black clouds and sharp wave crests. Instead, I stared at a black wall of solid rock. The *bams* and *clunks* were *Resolution*'s hull hitting stone.

But she wasn't aground. I pushed in the ignition key and turned it to start. I lock the transmission in gear when the engine is stopped to keep the propeller from turning, so when I started the motor up, it wound up to full speed ahead in an instant. *Resolution* surged away from the cliff, barging across the little inlet towards stone teeth that turned the waves to froth a hundred yards or so to starboard. If I didn't turn *Resolution*, she'd go on the rocks. Without even getting completely into the cockpit, I grabbed at the tiller to steer her clear of the rocks. I grabbed . . . and the tiller was there, but only suspended by the line and shock cord that had kept *Resolution* on course, and a broken end riding on the lower edge of the transom tiller opening. It wasn't attached to the rudder. One or more of the *bams* that woke me had snapped the tiller at the rudder head. I could only watch in horror as the little Bukh engine drove *Resolution* hard onto the low-lying rocks I'd hoped to steer away from.

Habit forced me to put the engine in neutral. *Resolution* listed to starboard and a wave lifted her farther onto the rocks. I saw one just under the forward shroud that I could nearly step off onto. I can do nothing without a tiller, I thought, and my neck is more important than the boat. I grabbed the Emergency Position Indicating Radio Beacon (EPIRB), a Personal Location Beacon (PLB) version, and unclipped the portable VHF from the cockpit coaming. With this equipment in my oilies' pockets, I abandoned ship, stepping out onto the rock and then swimming a couple of strokes to a boulder closer to shore. I climbed up on it and pulled out the radio.

MAYDAY. MAYDAY. This is the yacht *Resolution*. On the rocks. MAYDAY. MAYDAY.

No reply.

I sat a few minutes, then broadcast another MAYDAY call. Still no answer.

Resolution rested on the rocks, listing to starboard. And I realized I had left my grab-bag and my passport case on her. Should I go back aboard? Would it be foolhardy? Damn the wind. Damn the waves. Full speed ahead. I went back into the ocean and made my way to the boat. Grabbing the top lifeline, I tried to lift my leg up so that my knee rested on the bulwark. The cold made my muscles refuse to stretch and I couldn't do it. But I could reach the jib furler line, which had a tail because the jib was furled. I made a loop in the tail and used it as a stirrup to clamber aboard over the lifelines. The Bukh rumbled on, its control panel screaming an overheat warning.

In the cabin, the grab-bag and the passport case hung exactly where I had put them, and it took only a moment to collect them. *Resolution* rolled on an incoming wave. I took another moment to look for the sat-phone, but it had jumped from its box and disappeared amidst the clutter on the cabin sole. The motor chugged. The overheat warning screeched. Incongruent, I know, but I thought, why ruin a good engine? I couldn't help removing the companionway stair, pulling the cover from the engine, and stopping it. (Bukh left one vital part off the engine, which it now supplies automatically, that allows connection of a stop cable.) In the quiet, the grinding of

Resolution's hull on the rocks was like a hero grinding his teeth to keep from screaming as the villain tortured him. Perhaps I, who slept when I should have been in control, was that villain. I turned on the ship's radio for a last MAYDAY call and said I was abandoning ship.

Engine cover replaced and companionway stair affixed, I clambered into the cockpit with grab-bag and passport case looped to my left wrist. The water was higher. *Resolution* rocked with the waves. If I were to get off, it had to be now. Over the lifelines and onto the rock. A few strokes to another boulder. Scramble out of the water. Clamber up to perch on top and turn on the EPIRB.

The tide was coming in and the waves were definitely higher. *Resolution* rolled to starboard with incoming combers, to port as they receded. She moved shoreward with each series of waves. If those incoming waves tipped her completely onto her starboard side, I'd be tangled in the rigging at the least and bludgeoned by the mast at most. Ten yards or so away, another rock sat at the edge of the cliff. Steep. Twenty-five or 30 feet above water. The only nearby place where I'd be safe.

Back into the water. This time only to my waist as a footpath of submerged rocks led from my precarious perch to the larger, hopefully safer one. Fingers and toes. Knees. Scramble. Inch upward one toehold, one handhold at a time, until I could sit on top of the rock. The EPIRB beeped and flashed, assuring me it was communicating with satellites that were communicating with the Coast Guard that was communicating with rescue helicopters that would be overhead before long. Until then, I could only shiver and wait.

From my perch, I watched as rising water and incoming waves lifted *Resolution* and carried her shoreward, then dropped her on unyielding rocks. At first she hit them with the same bam-bam sound that woke me. Then came the snapping and crunching of wood as stone teeth chewed at the Fijian kauri of *Resolution*'s hull.

A hole opened in the stern hold. I could see it when *Resolution* rolled to port. My wetsuit floated near the shoreline, along with cockpit seat cushions. No doubt my fins and mask were lying amid the rocks of the bottom.

Dong. Dong.

A sound like the ringing of a bell. I scanned the cliff line, thinking perhaps someone ashore had noticed the wreck.

Dong. Dong. Do-o-ong.

Clear as any Japanese temple bell cast of finest bronze. Then I spotted a gray cylinder in a pool at the base of my rock. The empty tank for my kerosene heater had dropped through the hole in the hull and now, each time it struck a rock, it pealed out the death knell of my beautiful boat.

Dong. Dong. Dong. Dong.

In the dawn, the word daybreak took on new meaning. Boat break. Dream break. Heart break.

Mission 56309

At the Auckland Rescue Helicopter Trust facility at Mechanics Bay, Auckland, the duty team stood by in the ready room. The call that started Mission 56309 came from the Civil Aviation Authority National Rescue Coordination Center at 6:04 am.

Crewman Leon Ford answered the phone, motioning to the pilot and paramedic to gather round. He switched to the speaker phone.

"Satellites picked up a 406 EPIRP signal on the northeast side of Great Barrier Island," the voice said.

"Position?"

"South thirty six oh six. East one seventy five twenty six," NRCC said.

"We're on it."

The team geared up and set up the 600-pound No.3 winch on the Bolkov Kawasaki 117 helicopter. Pilot Lance Donnelly ran through the preflight while Paramedic Bruce Kerr and Crewman Ford double-checked medical supplies and equipment on board. Preparations complete, the three men donned their flight helmets and took their places in the aircraft.

Lance turned on the ignition and started the rotors turning at 6:25 a.m., 21 minutes after the phone rang. In rescue missions, time is of the essence. No one knows what the situation is on the ground at the accident site. The BK117 lifted off at 6:32 a.m., and made a beeline for the coordinates

NRCC had supplied. It looked like the accident was somewhere near Waikaro Point.

As the chopper neared Rakino (Aryd) Island, Lance began a search using a Becker radio direction finder and Bruce donned the night vision goggles to enhance his vision in the predawn gloom. Halfway between Aryd Island and Great Barrier Island, the Becker picked up a signal at 121.5 Hz, and soon after, Bruce caught the flash of a strobe light against the dark cliffs of Waikaro.

The first flyby positioned the survivor on a lone rock away from the main island. Two more flybys helped the crew determine just how to extract the survivor.

A yacht lay on the rocks about 20 meters from the base of the cliff. It was low in the water and rocked sluggishly back and forth on the incoming waves. The survivor sat on a rock some 10 meters above the waterline and about the same distance out from the cliff line. The crew decided he looked fit and okay to winch off the rock into a hovering helo. Lance took the chopper back over the cliff and landed it in a clearing. The crew removed a stretcher and set up for a winch rescue.

By 7:00 a.m., the helicopter was hovering over the survivor and Bruce the paramedic had begun his descent on the cable. Leon operated the cable lift.

I stood up as the man on the end of the cable reached the top of the rock. "Are you Charles Whipple?" Paramedic Bruce Kerr asked.

"Yes," I said.

"Back up."

I took a step back. Bruce slipped the harness over my head and settled it under my arms. Signaling Leon, Bruce wrapped his arms and legs around me, and the lift whined upward. The helicopter was as motionless as if it were parked on asphalt.

At the door of the helo, Leon reached out to hold me as I set my feet on the undercarriage and clambered into the chopper. Bruce followed, and Lance took the bird back to where they left the stretcher on top of the cliff. The chopper landed atop the cliff at 7:11 a.m. The crew reconfigured the seats and put the stretcher back in. Leon sat facing forward, I faced backward, and Bruce sat with his back against the side of the fuselage. Leon handed me a headset so I could participate in radio conversations.

"Do a flyby," crewman Leon said, "I want to get some footage." He poked his little video camera out the open door.

Resolution lay low in the water, listed to port, decks white in the early morning light and sails aglow in Storm Orange. At the peak of her mast, the tricolor navigation light still burned. *I am a ship*, it said. *Green to starboard, red to port. I live. I live. I am not dead. I am not . . . .*

The Sack

I landed at the Auckland Rescue Helicopter Trust facility at Mechanics Bay, Auckland, with the clothes on my back and the sack clutched in my fist. So. What's in the sack of existence I dumped out on the table? An EPIRB. A portable VHF. A passport, driver's license, some yen, some NZ dollars, a yacht master's certificate, pictures of the girl, daughters, sons-in-law, grandkids, Tomlin the dog—one credit card and a discount card from Burnsco, a ship chandlery in New Zealand. A wet set of foulies. A yellow T-shirt that says RESOLUTION Around the World. Ha. Non-slip deck shoes. Old jeans. Merino wool undies, *the better to keep you warm with my dear for its winter in the Southern Hemisphere*, and socks with a hole in the heel. Ship registration for a boat that no longer existed. Radio license for a radio that could make no more calls. A spark of life. Memories. Tears. Disappointment. Disillusion. A piece of waxed sail twine. Holes that fit all I left behind. Shame. Regret. A sense of worthlessness. A muesli bar. But still a flicker of hope. *I am not dead...*

Sayonara Mechanics Bay

Just before we landed at the Westpac Auckland Rescue Helicopter Trust facility at Mechanics Bay, Lance the pilot told me that the police would be waiting.

What!

The police will be waiting, he told me again. They'll want you to show your EPIRB and explain how it saved your skin today. They want everyone who goes out in the rough, ocean or mountain, to carry a PLB. Just the other day someone got lost out hiking. They didn't have a PLB so there was no pinpointing their position. Almost killed them.

It happened just as Lance said. TV cameras were there. Police were there. Scenes of me and my EPIRB, cut away to footage of *Resolution* dying on the rocks of Great Barrier Island, and then a close-up of the PLB with me saying it saved my life.

Only then was I allowed to go into the station and change my clothes. Remember, I swam my way to the rock from which the crew reeled me up into the helo. Of course I had nothing to change into, so the rescue squad lent me togs to use until I could buy some. They also gave me a ride to the hotel.

Passport and credit card let me check in. The rescue trust people left. First I called the girl to assure her I was uninjured. That's when I learned that the Australian authorities, which picked up the QME AccuSAT locater beacon first, had called the girl and asked her where I was, just in case the beacon had been accidently set off. So for some hours before I called, she knew that I had set off the emergency locator beacon, calling for help. "I'm safe," I said.

"Good," she replied.

My next phone call was to John Welsford. "Resolution is gone," I said, and told him what had happened.

"Don't beat yourself up," he said, but I did anyway. Whose fault was it that Resolution died on the rocks of Great Barrier Island?

Mine.

No one else.

Mine.

I slept.

Come morning, John was there. He'd driven up from Hamilton and was ready to get me set to go. There was a neighborhood store where I could get underwear and clothes, which I did, changed, and bundled the borrowed clothes up to return to the rescue trust.

One more incident in praise of New Zealand.

The one thing I didn't have in my grab bag was meds. At the time, I was taking hypertension meds and needed to get a new prescription. No problem. I just went to the nearest general clinic, without an appointment, and explained my situation. The receptionist sat me down and within less than a half hour, I was in the doctor's office. I told him my story. He input my name into the system with his desktop computer, and my New Zealand charts popped up on his screen. "Ah, yes. I see. Now, what do you need?"

I told him I would be leaving New Zealand shortly and a two week supply would be more than enough. Moments later, I had the prescription and was headed for the nearest pharmacy. Everyone treated in a New Zealand clinic or hospital has their charts in the nationwide medical system. It worked for me.

My grab bag contained my return ticket to Japan, so all I had to do was schedule a flight out and return to my former life. Right. I wondered if I would ever sail again.

"You learn more from losing than winning. You learn how to keep going."
--Morgan Wootten

Epilogue

Building Confidence

To quote John Welsford again: "By the time you finish building your boat, you have all the skills you wish you'd had when you started."

I built *Resolution*, an ocean-going Sundowner from John's drawing board. I built her while the plans were being drawn. Or, put another way, each new sheet of plans came just as the information on the last sheet had been digested and turned into a piece of *Resolution*. John could see what I was doing, and he reflected that in each new sheet of plans. For example, we started off assuming the main mast would be made of old, straight-grain Oregon pine. As it turned out, however, a thick-walled aluminum flagpole with an Oregon pine bottom piece performed well as *Resolution*'s mast. Among the flotsam at the wreck site a decade later was the Oregon pine topped by a length of aluminum pole. When you build something right, it stays right.

That said, here I am, a decade after the fact, just now putting pen to paper in a serious attempt to chronicle my affair with the lady called *Resolution*, who pounded herself to bits on the rock teeth of Great Barrier Island.

Resolution lived only a little while, but she'll sail forever in my heart. Her willingness to ghost along much faster than anyone imagined. Her ability to handle even strong storm conditions without causing her skipper undue distress. Her steady attitude at any point of sail. She was truly an ocean-going craft built for a purpose. And had her skipper been more adept and more cautious, she might still be sailing among the islands and across the oceans.

Resolution was a sweet boat. And I'm proud to say, "I built her."

Charles T. Whipple

The Hand of Fortune

Until March 31st, 2018, I didn't fully realize the hand of fortune in the wreck of *Resolution*.

I woke to the banging of her hull, or perhaps her rudder, on rock. With my head out the main hatch, I thought I saw a clear channel to open water, but with the broken tiller, which I did not notice until after I'd fired up the Buhk and put it in forward. Of course I couldn't steer and *Resolution* went on the rocks for sure.

How fortunate was I? Fortunate enough to be able to step from the forward deck onto a protruding rock. Fortunate enough to have the tail of the jib furler in the water where I could make a loop to act as stirrup to put my foot into to get back on board for my grab bag. Fortunate enough to have a crag of rock close enough for me to climb up and out of the way. Fortunate enough to have an ACCUSAT PLB EPIRB to connect me to the Auckland Emergency Helicopter Trust (via Australia, by the way).

Leon Kerr was the medic crewman who came down with the cable to winch me aboard the hovering Auckland Rescue helicopter. As they pulled me into the whirlybird's cabin, Kerr said, "Somebody up there likes you, mate."

Again, I didn't realize how much fortune held me by the hand until much later. Somewhat more than a year later, in fact.

The year 2009 marked the 50th anniversary of my high school graduation. I don't travel to the USA often, so I took the opportunity not only to see my old classmates, but also to stretch my legs, so to speak, and visit relatives and friends in faraway places such as Denver, Colorado, Dallas, Texas, and various towns in Arizona, of course. I spent many hours nearly motionless in the bellies of modern airliners and it all caught up with me one night in November 2009.

I sat at the kitchen table when I was hit by tunnel vision and shortness of breath, along with a few seconds of dizziness. I told the girl she'd better call an ambulance, which she did. By the time the ambulance arrived, the tunnel vision had disappeared and my breathing was back to normal. At least that's what I said. (I really do not like hospitals.)

The EMT guys said I should ride the ambulance to the emergency hospital. In fact, they carried me down the flight of nearly 50 steps that lead up to my house. In moments, I was on the gurney and the ambulance was on its sedate way to the gang of doctors and nurses that now waited for me to arrive.

The ambulance sidles into the emergency bay and the rear doors were jerked open. "You're at the hospital now. You'll be all right," a male nurse said.

I made no reply.

A young doctor hooked up the IV, stabbing me with his needle. I know something of needles, you see, because I once had to have blood drawn every hour on the hour for 24 hours. Don't know what they were looking for, but they found nothing. That was late November of 1960 and in those days the needles used looked like hollowed out ten-penny nails. The needles in Chiba were hardly a pin prick in comparison.

IV installed, they wheeled me into the CT room and had a look at my internal gizmos.

After the doctors watched the CT screen religiously, nodding and commenting in low tones, one came to the head of the gurney I'd been moved to. Speaking Japanese (of course) he said, "Your problem is . . ." I looked nonplussed at the ¥10,000 words he used.

"Better known as Economy Class Syndrome. In English, they say pulmonary embolism."

In layman's terms, blood clots had formed in the veins of my legs, broken loose, and made their way up through my heart and into my pulmonary arteries. Fortunately, the clots did not shut off all the blood to my lungs, so I am alive. Fortunately, as well, because of the death of *Resolution*, I was within easy reach of competent medical assistance instead of being alone aboard my boat somewhere on the Indian Ocean. Fortunate, indeed.

A Visit to the Grave

Fast forward to March 31st, 2018. Great Barrier Island is just a short flight from Auckland, but it's the dickens to get an empty seat. Maybe that's because twelve, including the pilot, is all the little puddle jumper can carry, and that one usually makes one round trip a day, sometimes two.

A couple of days before flying to the island, I called the Mabey residence. Isabel Mabey answered, "Hello."

"Hello," I said. "This is Charles Whipple speaking."

"Oh. YES! Yes." She knew immediately who I was, which surprised me considerably.

"John Welsford and I are flying over to Great Barrier on the 31st. Could we come and visit?"

"Oh, yes. Certainly. We'll be waiting for you."

And they were. Scott and Isabel Mabey. Owners of the Mabey farm. Raisers of prime Angus beef cattle.

Mike Newman met us at the airport and drove us to the Mabey farm while filling us in with a running commentary on things Great Barrier Island. New Caledonia claims to be close to heaven, but Great Barrier is only a step back.

Scott and Isabel were waiting at the farm house when we arrived. What a reception. It was my first meeting with these gentle people, but it was like greeting long lost friends. And we learned things there that I had not known before. Like, when GME, the Australian company that monitored my EPIRB, notified New Zealand authorities, those people called Scott Mabey. He jumped on his dirt bike and went to the heights overlooking the place where *Resolution* and I were rockbound. I'd noticed a motorcyclist, but thought it someone curious about why the helicopter was flying around there.

Some days later, Scott took his boat around to the wreck site. He rescued *Resolution*'s life ring, a porthole, my ukulele, and a Maori Tiki that had ridden shotgun on the boat. He put the life ring and porthole in the garage, but sent the ukulele and the Tiki to me in Japan, after contacting the authorities to find out who owned *Resolution*.

Imagine. Two weeks after the Auckland Helicopter Rescue Trust crew successfully reeled me up off the crag of rock I was on, a package arrived from New Zealand with a broken ukulele and a Maori Tiki in it. "We thought you'd like these," Isabel wrote. How heart wrenching can that be?

Despondent as I was from having allowed my ship of dreams to go on the rocks of Great Barrier Island, knowing that the ukulele—made for me by a diving buddy and signed by every diver in the Manatees Scuba Diving Club—had come back to me, and that the Maori Tiki set to accompany me around the world, was also both back in my possession gave my spirits a tremendous boost.

I sent the ukulele to the man who made it, and he returned a few weeks later, good as new.

The Tiki was not damaged beyond a little scraping. Polished with an oiled cloth, the Tiki stands proud, next to the refurbished ukulele on the cabinet in the vestibule of my home in Chiba. They are reminders of *Resolution*, my ship of dreams.

"Took my daughter out fishing," Scott said. "We caught a marlin. It's been in the smokehouse now long enough to be about done. Would you like some?"

Fresh smoked marlin? What a way to lunch. "That'd be nice," said John. All I could do was nod an emphatic agreement.

Moments later, we had thick slices of smoked marlin in front of us.

What a lunch!

"We'd like to go have a look at the *Resolution* wreck site," John said.

"Bit of a hike," Scott said. "You'll need to go down Bucklands Beach a ways. There's a white picket fence just a little this side of where the beach ends. You'll want to start climbing there. Once you top out and go around, I'll meet you there with my motorbike and show you where to go down to the wreck site.

Sounded simple.

Simple-sounding things to do never turn out to be simple. That should be another of John Welsford's theorems.

We made our way to the gate that stands between the Mabey place and Bucklands Beach. We went through the gate, careful to latch it again, and walked down the two-rut road to the beach. And what a beach it was. Flawless. Not marred by a single footprint as far as we could see. Far along the beach, however, a group of revelers cavorted about a couple of jet skis. Their power cruiser was anchored offshore.

As Scott said, near the end of the beach stood a white picket fence, one that surrounded a grave. Perhaps an ancestor. We took that as reason to start climbing the striated ridge that towered above the placid beach. Unfortunately, when I put too much physical strain on my heart, it goes to doing premature ventricle contractions (PVCs) about every three beats. While it's not a fatal condition, it is uncomfortable (worrying) to me. About two-thirds of the way up that ridge, I chickened out.

"Not gonna make it, John," I said.

"Right. Well. You stay here and I'll be back in an hour." He took off up the ridge with the stride of a man half his age.

An hour came and went.

John didn't come back. Who knows why?

I decided to continue up the ridge, over the edge, and down the other side, all the way to the water below. I could see then that the wreck site was not just on the far side of the first ridge, but up and over another larger ridge beyond that. But it looked like I might be able to get around the edge of the ridge where it came down to the water about half a mile ahead.

Clenching my teeth, I started down the ridge.

"Hey!" John was standing at the edge of the water below. "Easier way over there." He waved toward a part of the ridge with less of an angle. I followed his suggestion and soon stood beside him. He held a ragged bit of *Resolution*'s hull. "Thought you might want this."

"Thanks" was all I could say. That shard of *Resolution*'s hull came all the way back to Chiba with me, wrapped around my suitcase by a thoughtful Air New Zealand check-in officer.

So all I have of the boat I built with my own two hands, a few power tools, and the help of my friends in New Zealand is a refurbished ukulele, a Maori Tiki, a scrap of her hull, one porthole from her cabin, and the cork that popped from the sparkling apple juice used at her christening.

Still, as fortune and somebody up there are on my side, I live. I've lived to write this memoir of my ship of dreams— *Resolution*.

I will never forget her, nor will the dreams she afforded me fade. I hope you the reader can find both courage and assistance, and that you too may realize that after all is said and done, fortune holds the key, and pursuit of a dream is never in vain.

fin

Ship of Dreams

~ * ~ * ~ * ~

"It is not the critic who counts;
not the man who points out how the strong man stumbled or where the doer of deeds could have done them better.
The credit belongs to the man who is actually in the arena, whose face is marred by dust and sweat and blood; who strives valiantly;
who errs and comes short again and again;
who knows great enthusiasms,
the great devotions;
who spends himself in a worthy cause;
who at the best, knows in the end the triumph of high achievement,
and who, at the worst, if he fails, at least fails while daring greatly
so that his place shall never be with those timid souls who neither know victory nor defeat."
 --Theodore Roosevelt

~ * ~ * ~ * ~

About the Author

"The only thing I do well is write."
Charles T. Whipple is an international award-winning copywriter, journalist, author and novelist. His awards include Editor & Publisher Magazine DM Award, World Annual Report Competition Award, 2010 Oaxaca International Literature Award, and 2011 Global eBook Award.

Whipple was born in Show Low, Arizona. He spent two and a half years in Japan as a volunteer youth missionary, and majored in Japanese History as a graduate student and grantee at the East West Center, University of Hawaii. He is fluent in spoken and written Japanese, and has long been interested in the fantastic aspect of traditional Japanese tales.

Whipple lives in the city of Chiba, the capital of Chiba Prefecture, which encompasses the ancient Kanto Kingdoms of Awa, Kazusa, and Shimosa. Today, Chiba hosts the Magic Kingdom of Disneyland and is gateway to Japan via the international airport in Narita.

He has one wife, four daughters, two sons, 18 grandchildren, and one dog. Whipple also writes western novels under the pen name of Chuck Tyrell.

Visit Charlie at:
Blog: http://chucktyrell-outlawjournal.blogspot.com/
Smashwords Bio:
https://smashwords.com/profile/view/CTWhipple

Other writing by this author

FICTION
CHARLES T. WHIPPLE

Seeing Japan
A Matter of Tea
The Masacado Scrolls series
    The Fall of Awa – Story 1
    The Shadow Shield – Story 2
    The Road to Kio – Story 3
    The Horse Soldiers – Story 4
    Nami of the Waves – Story 5
    Woman With No Name – Story 6
    The Annointed – Story 7
Volume I – The Masacado Scrolls

CHUCK TYRELL

Vulture Gold
Revenge at Wolf Mountain
Trail of a Hard Man
Guns of Ponderosa
The Killing Trail
Hell Fire in Paradise
A Man Called Breed
The Snake Den
Lightning Strikes Twice
Return to Silver Creek
Pitchfork Justice
The Killing Trail
Dollar a Day
Diablo
A Man Called Breed

Road to Rimrock
Dead Man's Trail
Blessing

Stryker series
    Stryker's Law
    Stryker's Ambush
    Stryker's Bounty
    Stryker's Posse
    Stryker's Woman
    Stryker's Misfits I
    Stryker's Misfits II: Jolsanny's Raid

NON-FICTION
CHARLES T. WHIPPLE

(in Japanese)
    English from Words You Know
    Effective Business Letters
    English for Divers
(in English)
    Seeing Japan
    Inspired Shapes (translated into English)

Made in the USA
Las Vegas, NV
19 November 2020